Science Learning for **ALL**

Celebrating Cultural Diversity

An NSTA Press Journals Collection

NATIONAL SCIENCE TEACHERS ASSOCIATION

ARLINGTON, VIRGINIA

Shirley Watt Ireton, Director
Beth Daniels, Managing Editor
Judy Cusick, Associate Editor
Jessica Green, Assistant Editor
Linda Olliver, Cover Design

Art and Design
Linda Olliver, Director
NSTA Web
Tim Weber, Webmaster
Periodicals Publishing
Shelley Carey, Director
Printing and Production
Catherine Lorrain-Hale, Director
Publications Operations
Erin Miller, Manager
*sci***LINKS**
Tyson Brown, Manager

National Science Teachers Association
Gerald F. Wheeler, Executive Director
David Beacom, Publisher

NSTA Press, NSTA Journals, and the NSTA Website deliver high-quality resources for science educators.

Science Learning for All: Celebrating Cultural Diversity
NSTA Stock Number: PB156X
ISBN 0-87366-194-X
Library of Congress Control Number: 2001087924
Printed in the USA by IPC Communication Services
Printed on recycled paper

Copyright © 2001 by the National Science Teachers Association.
The mission of the National Science Teachers Assocation is to promote excellence and innovation in science teaching and learning for all.

Permission is granted in advance for reproduction for purpose of classroom or workshop instruction. To request permission for other uses, send specific requests to:

NSTA Press
1840 Wilson Boulevard
Arlington, Virginia 22201-3000
www.nsta.org

Contents

Acknowledgments v
NSTA Position Statement on Multicultural Science Education vi
Introduction vii
Correlations with the *National Science Education Standards* ix

Curriculum Reform

1 **Cultural Inclusion** 1
   Where does your program stand?
   H. Prentice Baptiste and Shirley Gholston Key (February 1996)

2 **Embracing Diversity** 4
   As the U.S. population becomes more culturally diverse, schools can
   profit from this rich heritage
   Gerry M. Madrazo, Jr. (March 1998)

3 **Encouraging Equitable Enrollment** 8
   One district's efforts toward increasing minority participation in
   math and science
   Stan Hill and Paul B. Hounshell (February 1997)

4 **Make the Curriculum Multicultural** 12
   Act now and make science an inclusive endeavor
   Napoleon A. Bryant, Jr. (February 1996)

5 **Inclusive Classrooms** 16
   A multicultural look at the *National Science Education Standards*
   Konstantinos Alexakos (March 2001)

6 **Inclusive Reform** 20
   Including all students in the science education reform movement
   Mary Monroe Atwater and Melody L. Brown (March 1999)

Teaching Strategies

7 **Creating a Culture for Success** 25
   Teachers accommodate different learning styles and boost achievement
   *Barbara S. Thomson, Mary Beth Carnate, Richard L. Frost, Eugenie W.
   Maxwell, and Tamara Garcia-Barbosa* (March 1999)

8 **Capitalizing on Diversity** 30
   Strategies for customizing your curriculum to meet the needs of all students
   Lenola Allen-Sommerville (February 1996)

9 **Big Picture Science** 34
   Uncovering teaching strategies for underrepresented groups
   Charlotte Behm (March 2001)

10 Multicultural Teaching Tips **38**
Practical suggestions for incorporating the diverse history of science into the classroom
S. Wali Abdi (February 1997)

11 Teaching Essentials Economically **42**
Using dedication, not frills, to encourage student achievement in science
Joy R. Dillard (March 2000)

12 Structured Observation **46**
Charting student-teacher interactions to ensure equity in the classroom
Ellen Johnson, Barbara Borleske, Susan Gleason, Bambi Bailey, and Kathryn Scantlebury (March 1998)

13 Notable Women **50**
Teaching students to value women's contributions to science
Cindy L.F. Zacks (March 1999)

Science and Language

14 Language Diversity and Science **54**
Science for limited English proficiency students
Elizabeth Bernhardt, Gretchen Hirsch, Annela Teemant, and Marisol Rodríguez-Muñoz (February 1996)

15 Meaningful Lessons **58**
All students benefit from integrating English with science
Alan Colburn and Jana Echevarria (March 1999)

16 Science as a Second Language **62**
Verbal interactive strategies help English language learners develop academic vocabulary
Carmen Simich-Dudgeon and Joy Egbert (March 2000)

17 Scientific Literacy for All **68**
Helping English language learners make sense of academic language
Cynthia Carlson (March 2000)

Appendices

Appendix A Author Contact Information **73**
Appendix B Publications **76**

Acknowledgments

The National Science Teachers Association (NSTA) is devoted to promoting excellence and innovation in science teaching and learning for all. *Science Learning for All: Celebrating Cultural Diversity*, which represents the best articles of five years of *The Science Teacher*'s multicultural issues, aims to help educators include all students in science education.

This collection was developed by *The Science Teacher* staff and the NSTA Committee on Multicultural/Equity in Science Education. Shelley Johnson Carey (Director, Periodicals Publishing, NSTA) and Janet Gerking (Editor, *The Science Teacher*) selected the articles. Committee members Connie Cook Fontenot (Bethune Science Academy, Houston, Texas), Ellen Gaylor (Iolani School, Honolulu, Hawaii), and Elizabeth Hays (School of Natural and Health Sciences, Barry University, Miami, Florida) reviewed the articles and offered valuable feedback about resources and connections to the *National Science Education Standards*.

Jessica Green was the project editor for *Science Learning for All: Celebrating Cultural Diversity*. Beth Daniels and Michelle Chovan also provided assistance and advice in developing this compendium. Linda Olliver designed the book and the cover, Nguyet Tran did book layout, and Catherine Lorrain-Hale coordinated production and printing of the book.

NSTA Position Statement on Multicultural Science Education

Preamble

Science educators value the contributions and uniqueness of children from all backgrounds. Members of the National Science Teachers Association (NSTA) are aware that a country's welfare is ultimately dependent upon the productivity of all of its people. Many institutions and organizations in our global, multicultural society play major roles in establishing environments in which unity in diversity flourishes. Members of the NSTA believe science literacy must be a major goal of science education institutions and agencies. We believe that ALL children can learn and be successful in science and our nation must cultivate and harvest the minds of all children and provide the resources to do so.

Rationale

If our nation is to maintain a position of international leadership in science education, NSTA must work with other professional organizations, institutions, corporations, and agencies to seek the resources required to ensure science teaching for all learners.

Declarations

For this to be achieved, NSTA adheres to the following tenets:

- Schools are to provide science education programs that nurture all children academically, physically, and in development of a positive self-concept;

- Children from all cultures are to have equitable access to quality science education experiences that enhance success and provide the knowledge and opportunities required for them to become successful participants in our democratic society;

- Curricular content must incorporate the contributions of many cultures to our knowledge of science;

- Science teachers are knowledgeable about and use culturally-related ways of learning and instructional practices;

- Science teachers have the responsibility to involve culturally-diverse children in science, technology and engineering career opportunities; and

- Instructional strategies selected for use with all children must recognize and respect differences students bring based on their cultures.

—*Adopted by the NSTA Board of Directors, July 2000*

Introduction

What is a "multicultural" classroom? Classrooms, even if they are filled with non-majority students, are not necessarily multicultural. There are three elements necessary for a truly multicultural science-learning environment: first, the sense that all students can learn and do science; second, the view that each student has a worthwhile place in the science classroom; and third, an appreciation for the contributions of all cultures to our scientific knowledge (Atwater, 1993; Hays, 2001). *Science Learning for All: Celebrating Cultural Diversity* focuses on the need for multicultural science classrooms, and addresses what makes a culturally diverse science classroom a multicultural one.

The last two decades have seen increasing interest devoted to multicultural education. The National Science Teachers Association (NSTA) recognizes its responsibility and duty to promote quality science education to diverse student populations, and states that NSTA's mission is "to promote excellence and innovation in science teaching and learning for all." In the early 1970s, minority NSTA educators began meeting informally. About 1980, these NSTA members formed a group to address concerns of minority educators, equity in science education, and the growing interest in multicultural science education. This group—originally named the Black Caucus but now known as the Minority Caucus—continues to serve as an important avenue within NSTA for voicing concerns about multicultural science education.

In 1989, NSTA formed the Multicultural/Equity in Science Education Committee to review NSTA policies, programs, and activities relating to multicultural science education. The Committee developed the NSTA position statement on multicultural science education in 1991 and revised the statement in 2000 (see page vi). The Committee often solicits feedback from an associated group of the Minority Caucus, the Association of Multicultural Science Educators (AMSE). The two groups co-sponsor Share-a-Thons at NSTA conventions, where teachers share their lesson plans, ideas, and resources. The Multicultural/Equity Committee also reviewed the articles chosen for this collection and guided the development of this book.

Each year *The Science Teacher*, NSTA's journal for high school educators, devotes an issue to multiculturalism in science education. *Celebrating Cultural Diversity* represents the best of these multicultural issues of *The Science Teacher*. One of the articles in this book holds a personal connection for me. When I read "Teaching Essentials Economically," I was very impressed with author Joy Dillard, a young minority educator making a difference in a rural African American science classroom in an impoverished area. This teacher was modeling success against all odds. I wrote to her to express my feelings about her article and encouraged her to persevere because she was doing good science education in spite of her school's isolation and lack of resources. At my urging, and with the help of an AMSE mentor, Joy is becoming more involved with NSTA and AMSE to learn about and guide multicultural science education. Joy Dillard epitomizes the hard work and dedication of many science teachers striving to make a difference in their classrooms. It is for teachers like her that *Celebrating Cultural Diversity* was compiled.

The 17 articles in this book have been grouped into three sections: Curriculum Reform, Teaching Strategies, and Science and Language. Each article has also been reviewed to determine which of the *National Science Education Standards* are addressed (see Figure 1).

The Curriculum Reform articles help teachers evaluate the cultural inclusiveness of their science curricula. Three of the articles describe how and why schools can profit from the diverse heritage of the United States by developing a learning environment of understanding, respect, and multiculturalism. Educators are urged to emphasize student unity, equity, and diversity. The other articles give examples of how to change school policies, programs, and curricula to include all students in science education and to meet the learning needs of all students.

The Teaching Strategies section highlights techniques that teachers can adapt for their individual classrooms. Teachers learn to employ cultural knowledge, sensitivity, and interpersonal skills to maximize students' learning potential. The articles showcase innovative teaching programs that improve learning in diverse classrooms. This section also offers concrete teaching tips and practical suggestions for incorporating the diverse history of science into the classroom.

The final section, Science and Language, addresses the special concerns of learning the language of science while also learning English. The articles provide practical suggestions for integrating science with English, where English is a student's second language. Science itself can also be seen as a second language, and the articles offer methods to help students acquire academic language and scientific vocabulary. All students—English language learners and those already fluent—benefit from a curriculum that emphasizes the in-depth teaching of concepts, process skills, and critical-thinking skills. When I teach science to non-science majors at the college level, I emphasize that part of being successful in science courses is learning the language of the field, and I suggest that students treat my course as they would a language course. This helps some students overcome the anxiety of science that is often brought to the college-science classroom.

Celebrating Cultural Diversity provides valuable ideas and strategies for bringing multicultural education to your classroom. The articles demonstrate that such an equitable classroom is inclusive, provides opportunities for all students to learn science, and shows students that people like them can make scientific contributions to improve our world. The reader should recognize that this book is a collection from one source, *The Science Teacher*, and reflects the articles written for this journal. There is much more that has been published and said about multicultural science education. Appendix B lists additional resources suggested by the Multicultural/Equity Committee; this collection represents just a small number of the resources available to help teachers meet the needs of the diverse student population in their classrooms. Enjoy these articles and resources, and make your classroom a true multicultural learning environment.

Elizabeth Hays, Director
NSTA Multicultural/Equity in Science Education Committee

FIGURE 1.

Correlations with the *National Science Education Standards*

	Standard	Curriculum Reform						Teaching Strategies							Science & Language			
		1	2	3	4	5	6	7	8	9	10	11	12	13	14	15	16	17
Teaching	A Plan an inquiry-based science program for students.	•					•											•
	B Guide and facilitate learning.			•	•	•	•	•	•	•	•	•	•		•	•	•	•
	C Engage in ongoing assessment of teaching and of student learning.	•	•	•	•	•	•	•	•	•	•	•	•	•	•	•	•	•
	D Design and manage learning environments that provide students with the time, space, and resources needed for learning science.		•		•	•		•	•	•	•	•	•		•	•	•	
	E Develop communities of science learners that reflect the intellectual rigor of scientific inquiry and the attitudes and social values conducive to science learning.	•	•	•	•	•	•	•	•	•	•	•	•	•		•	•	•
Professional Development	A Learning essential science content through the perspectives and methods of inquiry.						•	•	•	•	•	•	•				•	
	B Integrating knowledge of science, learning, pedagogy, and students; it also requires applying that knowledge to science teaching.	•	•					•	•	•	•	•		•		•		•
Assessment	A Assessments must be consistent with the decisions they are designed to inform.					•		•	•		•	•	•	•				
	B Achievement and opportunity to learn science must be assessed.					•					•	•	•	•	•	•	•	
	D Assessment practices must be fair.					•		•	•		•	•		•	•	•	•	

FIGURE 1 CONTINUED.

Correlations with the *National Science Education Standards*

	Standard	Curriculum Reform						Teaching Strategies							Science & Language			
		1	2	3	4	5	6	7	8	9	10	11	12	13	14	15	16	17
	B The program of study in science for all students should be developmentally appropriate, interesting, and relevant to students' lives; emphasize student understanding through inquiry; and be connected with other school subjects.	♦	♦	♦	♦	♦	♦	♦	♦	♦	♦	♦		♦	♦	♦	♦	♦
Program	**E** All students in the K–12 science program must have equitable access to opportunities to achieve the *Standards*.	♦	♦	♦	♦	♦	♦	♦	♦	♦	♦	♦	♦		♦	♦	♦	
	F Schools must work as communities that encourage, support, and sustain teachers as they implement an effective science program.		♦	♦	♦		♦	♦	♦	♦								
System	**E** Science education policies must be equitable.	♦						♦			♦		♦			♦	♦	
	F All policy instruments must be reviewed for possible unintended effects on the classroom practice of science education.				♦		♦	♦			♦						♦	

Cultural Inclusion

Where does your program stand?

SEPTEMBER 1995: A MEXICAN FAMILY just moved to a small Midwestern city where the mother had been recruited to be a professor at the state university. The father was employed as a teacher in a local school district. The mother accompanied her daughter, a sophomore, to the local high school for registration and counseling. The daughter was excited about the prospect of registering for a genetics course and an advanced biology course because of the excellent science experiences she had the previous year. The counselor ignored the daughter's desire to take the advanced biology courses and told her, "I believe you will be happier in the regular biology course because most girls are." The young lady has been bored because she has had much of the work previously.

April 1993: A lesson on the circulatory system was being conducted in a middle school classroom. As a part of discussion on blood transfusion and preservation, the contributions of Charles Drew were included. During this discussion the science instructor passed out photographs of Drew. An African American student in the class, on seeing the photos, exclaimed "Was Dr. Drew Black?" The instructor responded "Yes, he was." The African American student replied "He *couldn't* have been (Black) and *done* all those good things."

August 13, 1995: *Parade* magazine printed a short feature entitled "Recalling the Golden Age of Inventors," which focused on inventions created in the United States from 1850 to 1900. Several inventors—Samuel Morse, Thomas Edison, Elias Howe, Alexander Graham Bell—all white males, were depicted as receiving help from the *Scientific American* magazine patent office in applying for patents for their inventions. However, there was no mention of Jan Matzeliger, Elijah McCoy, Granville T. Woods, Andrew Jackson Beard, and Lewis Latimer, all well-known African American inventors of this period.

These examples illustrate the need for implementing science that is multicultural. Science should be presented in a non-elitist, culturally diverse context that includes all students. Presenting science as a hands-on, activity-based, problem-solving instructional program couched in constructivist theory enables all students, including students of color, to excel in science.

When African American students were asked through a survey and interviews (Key, 1995) whether they prefer to study science topics relevant to their culture, they responded positively. Examples of potential science topics included in the survey were:

- Benefits of African American astronauts in space flights,
- The inventions of African American scientists and engineers,
- How African American inventions have affected society,
- The diseases of African Americans such as sickle-cell anemia,
- The diseases that affect Hispanic Americans, and
- The diseases that affect Asian Americans.

These kinds of culturally inclusive items were chosen at a much higher rate ($p < .001$) by students of color than items on the survey that did not specifically address their cultures.

TYPOLOGY FOR CULTURAL INCLUSION

The following typology can help science teachers evaluate and determine to what extent cultural inclusion is present in their science program. Cultural inclusion is

BY H. PRENTICE BAPTISTE AND SHIRLEY GHOLSTON KEY

FIGURE 1.

Typology for multiculturalizing science.

Level III: Process/philosophical orientation. Science teachers must be social activists. They can help their science students promote equal opportunity, respect for those who differ, and power/equity among groups both in the school and in the local community. After teachers have internalized the parameters of Levels I and II, they actively design, develop, or seek out science programs that are truly antiracist and multicultural. Science instruction is then culturally inclusive with a commitment to the philosophy of multicultural science education.

Level II: Process/product. Infusion into science curriculum of diverse cultural perspectives and contributions to a science concept development and/or evolvement. Scientists of color and women scientist's experiences and contributions are tied into the teaching of science concepts and topics. Science instruction is commensurate with diverse learning and cognitive styles. Problem solving processes and scientific method are used in elucidating the faultiness of racial and sexual stereotypes, prejudices, and ethnocentrism. Contextual array of science content is permeated with diversity.

Level I: Product. Focus on scientists of color and women scientist's contributions, in isolation. Highlighting an ethnic or cultural group invention, discovery or contribution, or scientist of color birthday. Multicultural additions are made to the curriculum in merely an additive, superficial way. A science textbook might end each chapter with a vignette on a scientist of color. During African American History month, students might read isolated material about famous African Americans in the field of science.

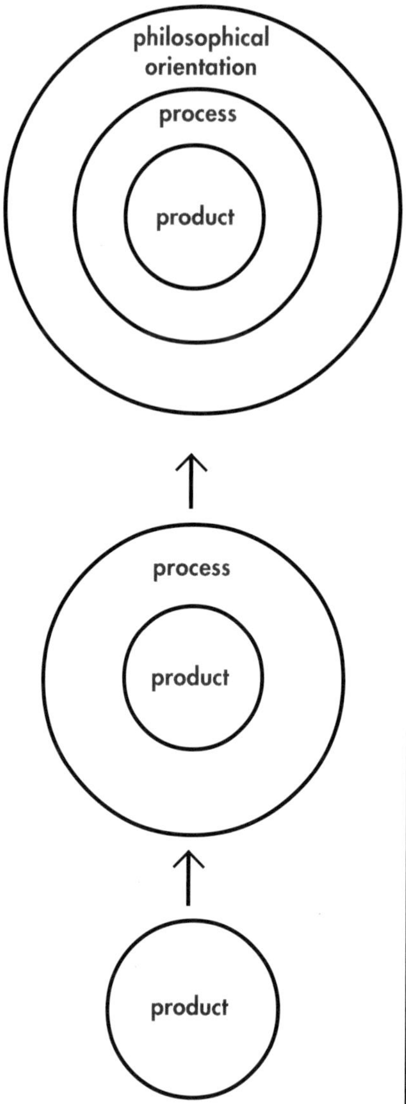

The typology of cultural inclusion (See Figure 1) is modeled after Baptiste's (1994, 1992, 1986) typology for assessing multicultural science education. The model has three levels, each encapsulating the previous one or ones. The model enables science teachers to determine the extent that currently used instructional strategies, curriculum, textbooks and other resources, activities and laboratory experiences reflect an internalization of cultural inclusion. Each level has a distinct set of characteristics and represents a qualitative level of accomplishing cultural inclusion.

LEVEL I

This level can be described as additive and tangible. Science experiences, perspectives, and contributions of people of color and women are presented in isolation or as additives to the regular science curriculum at Level I. Among science teachers, as with other teachers, this is the most popular strategy for multiculturalizing science instruction. The following topics are illustrative of Level I.

■ Studying African American scientist Charles Drew's work on blood preservation and organization of the first blood banks in a biology class during Black History month as opposed to incorporating his scientific contributions into the curriculum lessons on blood and circulation. Studying contributions of the Mexican American physicist Luis W. Alvarez (winner of 1968 Nobel Prize in Physics) during May (Cinco de Mayo) or September (Mexico independence) as opposed to studying his contributions during the curriculum focus on subatomic or resonance particles.

■ As part of the school's celebration of Native Americans' contributions, the science teachers use some of the following contributions and experiences of Native Americans: the Anasazi recorded the supernova of A.D. 1054 in paintings and built solar observatories in what is now Arizona and New Mexico; the Zuni people's historical contributions to ecological, engineering and land man-

the integration of the learner's culture into the academic and social context of the science classroom to aid and support academic learning (Key, 1995). It values cultural identity while promoting personal, human, and social development. Cultural inclusion is needed to develop competent science students who are socially responsible participants of a culturally diverse society where group identity is valued and preserved.

Cultural inclusion stresses changing science programs to make them culturally consistent, relevant, and meaningful to diverse populations. It helps to eliminate bias, to create a new standard of measure, and to provide equitable curriculum and pedagogical practices.

agement of the arid lands that they have occupied for centuries; the practice of agricultural polycropping (intercropping) facilitated soil enhancement in nutrients and a prevention of soil erosion, and numerous contributions of medicinal derivatives (such as quinine) from indigenous plants for the treatment and prevention of certain diseases.

LEVEL II

There is an infusion into science curricula (Figure 1) of diverse cultural perspectives and contributions to a science concept development and/or evolvement. Contributions of scientists of color and women are integrated into the science curriculum, not just tacked on. Science teachers operating at this level enable their students to use science problem-solving processes and the scientific method in elucidating the faultiness of racial and sexual stereotypes, prejudices, and ethnocentric behavior. The following examples are illustrative of Level II.

Ecological Technology. Students investigate products from "useless" raw materials. The peanut, cotton seed, and rice hull can be selected as raw products to investigate in both biology and chemistry classes. The discoveries and contributions of the following scientists: George W. Carver (peanuts), Eli Whitney (cotton seeds), Oto Yakamoti (rice hulls) include butter, oil, and fertilizer. Biology and chemistry students can study the lives of an ethnically diverse group of scientists as they replicate their investigations.

Cellular Biology. The study of the cell, the basic unit of life, takes place in life science, biology, and advanced biology classrooms; however, little mention is made of Ernest Just, an African American biologist who spent a lifetime studying cytoplasm, the cell membrane, and other components of the cell. This would be the appropriate time for the biology teacher to include Just and his contributions.

Earth Sciences. The study of the solar system and the universe tends to take on a Western cultural perspective in many of our Earth science classes. We very seldom, if at all, include in our Earth science classes Eastern perspectives and discoveries such as Zhou Yue, who around 300 B.C. in China modeled the Moon's movements; Hypatia, a woman mathematician in Alexandria, Egypt, who around A.D. 400 developed the astrolabe, an instrument used to observe the position of celestial bodies; Aryabhata the First of India, who in A.D. 497 deduced the Earth's rotation.

The manner in which the Islamic culture unites art, religion, and science into a fundamental world view is a perspective seldom shared in science classrooms. In spite of the Native American culture geographically within our midst, our ethnocentric attitudes have prevented us from understanding their sophisticated perspective of the complex relationship of science, religion, and nature that provides their world view.

LEVEL III

Science teachers operating at Level III of Baptiste's typology for multiculturalizing science are social activists. His or her science instruction is culturally inclusive. The science teacher at this level is committed to design, develop, or seek out science programs that are truly antiracist and multicultural. This science teacher will help students promote equal opportunity, respect for those who differ, and promote power equity among groups both in the school and in the community.

Merely presenting illustrative examples of science lessons reflecting Level III will not suffice. Level III has a strong philosophical orientation couched in the fundamental principles of equity, valuing of the respect for human diversity, and moral commitment to social justice for all. An equitable learning environment must be established in the classroom that positively supports various learning styles and all science instruction and content must be purged of all elitism.

Multiculturalizing science must focus on pedagogy and content concerns. Science teachers operating at Level III of Baptiste's typology understand the importance of implementing instructional strategies that are amenable to all students and are culturally diverse; and selecting science content that reflects various cultural perspectives and represents the diverse cultural roots of science knowledge. ✧

H. Prentice Baptiste is professor and associate director of the Center for Science Education, Kansas State University, Manhattan, KS 66506-5313. Shirley Gholston Key is visiting assistant professor and director of elementary education, University of Houston–Downtown, One Main Street, Suite 1075-N, Houston, TX 77002.

REFERENCES

Baptiste, H.P., Jr. 1994. The multicultural environment of schools: Implications to leaders. In *The Principal as Leader,* Larry Hughes, ed. (pp. 89-109) New York: Macmillan College Publishing.

Baptiste, H.P., Jr. 1992. Multicultural education: Its meaning for science teachers. In *Science Matters: A staff development series,* New York: Macmillan/McGraw-Hill.

Baptiste, H.P., Jr. 1986. Multicultural education and urban schools from a sociohistorical perspective: Internalizing multiculturalism. *Journal of Educational Equity and Leadership* 6(4): 295-312.

Key, S. 1995. African-American eighth grade students perceived interest in topics taught in traditional and nontraditional science curricula. Unpublished dissertation.

EMBRACING

As the U.S. population becomes more culturally diverse, schools can profit from this rich heritage

DIVERSITY

THE FACE OF THE AMERICAN CLASSROOM has changed dramatically in recent years and will continue to do so. As the number of students from varied ethnic backgrounds grows, so will the mix of ideas, interests, languages, and cultures brought to the classroom. This mix offers teachers an exciting opportunity to provide insight and leadership while integrating the diversity of their students' backgrounds into their lesson plans.

Multicultural education means defining school goals so that all students have an equal opportunity to learn. The *National Science Education Standards* (National Research Council, 1996) are clear in their intent—science is for all students, regardless of age, gender, cultural or ethnic background, ability, aspirations, or interest and motivation in science.

In planning and implementing effective educational programs, schools should recognize that "ethnic identity and cultural, social, and economic background of students are as vital as [students'] physical, psychological, and intellectual capabilities," according to Geneva Gay (1994), professor of education at the University of Washington and an expert in cultural diversity and equitable education.

MULTICULTURAL CURRICULA

Integrating multiculturalism into any subject can be tricky, and science and math educators (who are thought to work with purely objective disciplines) may find it especially difficult to develop new frameworks. The process of integration can be illustrated in the form of a continuum (Figure 1) that begins with an additive process—adding something multicultural to science curriculum and instruction. Teachers can move from

BY GERRY M. MADRAZO, JR.

addition to integration to accumulation and then towards attainment of multiculturalism, called advocacy of multicultural science education.

The extent to which teachers and students relate data and information from various cultures to the concepts and theories of science is called integration. Teachers help students understand how knowledge is constructed. In many respects this process reflects the procedures by which various scientists create knowledge in their disciplines. How knowledge is constructed can be influenced by cultural factors, and the teaching and learning process is only enhanced by using one's own cultural knowledge and perspectives. This mode of constructing knowledge is called accumulation.

When teachers and students discuss the ways in which various frames of references and cultural assumptions influence the accumulation of knowledge they experience a multicultural perspective. This leads to transforming knowledge or advocacy. Teachers and students are empowered by the knowledge, attitudes, and skills necessary to survive in a multicultural society. They have attained a mode called advocacy.

In 1991, the National Science Teachers Association (NSTA) issued a position statement on multicultural science education that said the welfare of the American classroom is ultimately dependent on the productivity and general welfare of all students. NSTA outlined the responsibility of each level of the educational community for multicultural education and pointed out that the entire educational enterprise—educators, parents, industry, community leaders, and policymakers—must believe that all students can learn successfully and must be willing to commit resources toward this end.

School districts must design curricula and instruction to reflect and incorporate diversity, while individual schools must provide science education programs that nurture all children academically and help them develop a positive self-concept. Science teachers are encouraged to educate themselves about students' learning styles (which may be culturally influenced) and make all students aware of career opportunities in science, mathematics, and related technological fields.

As schools enter the 21st century and the U.S. population becomes more diverse than ever before, multicultural science classrooms should reflect the principles outlined below:

▲ A multicultural curriculum results in respect for diversity flowing from knowledge. With that respect will come the ability of people to live and work together in a diverse society.
▲ Teachers can help students develop the decision-making, problem-solving, social, and political skills necessary for participation in a culturally diverse society. (Teachers can encourage open discussion of how learners feel about the subject.)
▲ Laboratory investigations and course content can be integrated into authentic activities relevant to minority students' everyday lives, interests, and experiences.
▲ Science instruction should represent a variety of traditional and historical viewpoints that integrate literature, math, history, and the arts. By presenting science as an ongoing, creative story with many parts, students will see their own cultural experiences reflected in the lesson.

The content and methodology of multicultural science and curricula (including resource materials) should be significant to students in school and at home. The curriculum should help students see the connection between their local and global environments and think conscientiously and critically about their role in these relationships.

THE MULTICULTURAL ENVIRONMENT

Every science teacher integrates multiculturalism into education to a degree. The direction of multicultural science curriculum, teaching, and learning will be defined and acted upon by each teacher. The following strategies may be helpful to science teachers:

▲ Choose curricula and science programs that are culturally sensitive to diverse student populations, and particularly to those who are traditionally underrepresented in science.

FIGURE 1.

Continuum illustrating the integration process.

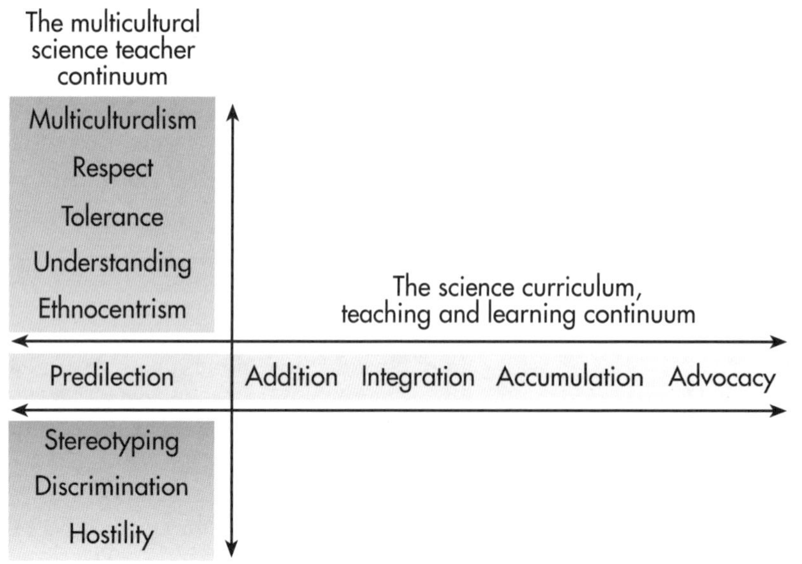

The multicultural science teacher continuum

Multiculturalism
Respect
Tolerance
Understanding
Ethnocentrism

Predilection

Stereotyping
Discrimination
Hostility

The science curriculum, teaching and learning continuum

Addition Integration Accumulation Advocacy

[Madrazo, 1997 (Modified from Hoopes, 1979; Banks, 1993)]

▲ Infuse discussions of scientific concepts and experiences with appreciation of the different cultures that influenced the nature and structure of the scientific enterprise.

▲ Maintain a classroom climate that encourages students to pursue careers in science, mathematics, medicine, engineering, and technology. Such a learning environment should reflect the equitable contributions of various scientists and educators.

▲ Use both cooperative and individual learning activities in doing laboratory investigations and during class discussions. Peer tutoring and problem-solving groups are especially useful and encourage students with different learning styles and backgrounds.

▲ Encourage students to be active participants in the learning process. Multicultural science instruction emphasizes dynamic inquiry and exploration, not static, memorized right and wrong answers.

▲ When discussing a lesson, examples can be used that appeal to all students. Teachers can incorporate student opinions into discussions to validate student understanding of concepts.

▲ Teachers can give students role models by bringing minority scientists into the classroom to talk about science and their field of expertise.

MYTHS AND REALITIES

Though many people recognize the need to address different viewpoints, some fear that without an agreed-upon cultural position, the classroom will become an arena in which ethnic groups clash and significant contributions of one group are reduced in order to laud the achievements of another. The same people fear that equity issues may be compromised. When carried out according to its true spirit, however, multiculturalism benefits all students as they gain an understanding of themselves and an appreciation for their peers. According to James Banks, an expert in multicultural education, the tremendous social and racial chasms that exist in American society make it naïve for anyone to believe that the study of ethnic diversity will threaten or exacerbate any notions of national cohesion (Banks, 1993).

Another common myth some educators believe is that "white" American or Western culture is excluded in multicultural studies. Contrary to popular belief, a multicultural curriculum does not leave out or undermine the European experience. Gary Howard, a multicultural education specialist who focuses on curricular and staff development, encourages white Americans to learn about other cultures and investigate what those cultures bring to the learning table. He also says that white Americans are not without their own cultural histories, though they may be several generations removed from those ancestors who "worked so hard to dismantle their European identity in favor of what they perceived to be the American ideal" (Howard, 1993).

Science teachers have always been at the cutting edge of changes in education and society. If we are to achieve scientific literacy for all students, science teachers and other educators must realize that curricula, as well as science teaching and learning, must change to reflect the diversity in our society. The better prepared our teachers are to embrace diversity, the more skilled our students will be to work, live, and prosper in a global community. ✧

Gerry M. Madrazo, Jr. is a clinical associate professor of science education and Executive Director of the Mathematics and Science Education Network at the University of North Carolina at Chapel Hill, 134 East Franklin Street, Chapel Hill, NC 27599-3345; e-mail: gmadrazo@email.unc.edu.

REFERENCES

Banks, J.A. 1993. Multicultural education: Development, dimensions, and challenges. *Phi Delta Kappan* 75(1):22-28.

Gay, G. 1994. At the essence of learning: Multicultural education. *Kappa Delta Pi Biennial.*

Howard, G.R. 1993. Whites in multicultural education: Rethinking our role. *Phi Delta Kappan* 75(1):36-41.

National Research Council. 1996. *National Science Education Standards.* Washington, D.C.: National Academy Press.

ENCOURAGING Equitable Enrollment

One district's efforts toward increasing minority participation in math and science

EDUCATORS CONSTANTLY TALK ABOUT the need for reform in science and mathematics education. We implement program after program and try innovation after innovation. Is it possible that we are looking in the wrong places, working on the wrong things?

What if the most essential place to focus is on the relationships within the school and school community? This includes the quality of the interactions between students and teachers, teachers and administrators, parents and school personnel, and business and the school family. Until we work at making these relationships more meaningful, alternative training plans, innovative hands-on programs, and well-meaning reforms will rest atop an unstable base.

As educators we always seek exceptional results and miraculous improvements, but we often wait for others to change. We want students to come better prepared, parents to be more involved, administrators to be more supportive, and legislators to provide more funding. Educators can work on all these areas, and we probably should, but until we choose to develop and transform ourselves, our goals for systemic change will remain unfulfilled.

The Winston-Salem/Forsyth County School System is an urban school district in the piedmont section of North Carolina. The system's minority population is approximately 42 percent, with African American being the dominant minority.

The system recently finished the first year of a five-year National Science Foundation grant funded by the Comprehensive Partnership for Math and Science Achievement (CPMSA). The grant is designed to increase the number of minority students who enter and successfully complete upper-level science and math courses. The specific benchmarks focus on:

1. The number of minority students completing Biology, Chemistry, and Physics.
2. The number of minority students completing Algebra I, Geometry, Trigonometry, Pre-Calculus, and Calculus.

The initiative is called Project JUST (Join Underrepresented in Science and Technology). The major goal is to create an atmosphere of systemic change within the district that results in minority students excelling in upper level math and science courses.

We are framing our reform efforts around the six National Science Foundation systemic change drivers:

1. How do systemic policies affect the project's goals and objectives?
2. How does district leadership, governance, and management affect the project's goals and objectives?
3. Does the district have a standards-based curriculum that is related to actual teaching practices?
4. How are periodic and annual assessments used to inform the district teachers, counselors, and administrators about their role in systemic reform?
5. Are professional development activities ongoing, developmental, content-based, and constructionally oriented?
6. Are partnerships, parental involvement, and public awareness active components within the project?

The focus of this article will be our district's efforts around drivers 2 and 3—the districts' leadership and

BY STAN HILL AND PAUL B. HOUNSHELL

instruction efforts—and how we can grow and develop in these areas.

LEADERSHIP, GOVERNANCE, AND MANAGEMENT

The district superintendent, Donald L. Martin, Jr., is the principal investigator for the CPMSA grant. He has aligned the goals of the project with the overall goals of the system and is committed to bridging the gap in academic achievement between white and minority students. This commitment is evident in every agenda he sets dealing with student performance—whether it is in the community, with his management team, or with the district's school board. As project director, I meet regularly with school staffs to emphasize project goals and promote systemic change.

Many of the major systemic interventions associated with Project JUST were created at the school level by teachers, counselors, and administrators. When the project first began, I met with each school's counseling and administrative staff. Meetings were also held with the math and science teachers within the school. Each school was invited to create a local intervention that would drive systemic changes. This effort has yielded some of our most impressive efforts:

■ A bridge course between Mineral Springs Elementary and Mineral Springs Middle schools allows fifth-grade students and teachers to work with sixth-grade middle school teachers in the middle school classrooms. This bridge occurs during the periods of intersession. (Mineral Springs Elementary is on a year-round schedule.)

- Walkertown Middle School trains its entire core teaching staff in a yearlong training initiative called Problem-Based Learning. This effort focuses on a partnership with the Bowman Gray School of Medicine of Wake Forest University.
- Parkland High School has created the Achievers Program, which allows students in biology and geometry to extend the time it takes for them to complete their course work as long as they are making satisfactory progress.
- Glenn High School has developed a training effort around effective communication. The local teacher association representative and I lead a staff development initiative designed to improve each teacher's ability to relate to their students as exceptional learners.
- Reynolds High School has created the district's first counselor-driven intervention. The counseling staff has provided special training sessions for parents, tutoring sessions for students, and evening classes in which community leaders work with students and parents on effective study skills, skills needed for the workplace, and the steps necessary to enter careers related to math, science, and engineering.
- North Forsyth High School has created both a summer bridge program and an in-school mentoring program. The bridge program places 9th- and 10th-grade students in a two-week enrichment session. Students improve their process skills and visit local businesses and industries where they are trained in the practical applications of the science and math concepts they are learning. The mentoring program is conducted by the Alpha Phi Alpha fraternity. Mentors work with students during the school day and provide after-school tutoring and enrichment opportunities. In addition, they monitor students' attendance, course selection, grades, and attitudes.

The district is combining many of its federal and private funds behind the CPMSA systemic reform goals and objectives. Eisenhower professional development funds, Howard Hughes Foundation funds, and local funds support the school initiatives mentioned above along with curriculum innovations.

The district is also partnering, through subcontractual arrangements and informed agreements, with other agencies, including Winston-Salem State University, Wake Forest University, Bowman Gray School of Medicine, the Alpha Phi Alpha Fraternity, the North Carolina Math and Science Educational Network, the North Carolina State Department of Public Instruction, the Winston-Salem Urban League, SciWorks Nature Museum, and R. J. Reynolds Company.

A STANDARDS-BASED CURRICULUM

The national standards in science and mathematics are the foundation for our standards-based curriculum. Teachers have worked through the summer to translate national goals and objectives into local goals and objectives that can be adopted and applied in actual teaching situations. We are emphasizing assessment, the teaching and learning environment, technology, effective communication, and professional teacher development.

A major effort in the area of curriculum and instruction focuses on problem-based learning. At its core, problem-based learning involves students in resolving real-world problems that require the acquisition, analysis, and interpretation of data and the use of problem-solving skills. This makes mathematics and science more relevant and allows students to work in cooperative groups and become investigators to solve problems. Students become the focus of the learning with the teacher guiding their inquiry.

The problems that students tackle (see example problem below) are jointly developed by teams of district math and science teachers and faculty members from the Bowman Gray School of Medicine. Topics explored include forensic science, bacteriology, the mathematics

---

Family Vacation
(A Problem-Based Learning Initiative)

Problem: Students pretend that they are just starting a new year of school and have been asked to write an essay about a memorable event or experience they had during the summer. Their task is to make up the ending to an essay about an unforgettable camping trip with friends from New York. On the camping trip, a six-year-old girl gets very sick.

Rationale/Goal: Seventh- and eighth-grade students will use the scientific method to study the concepts of risk exposure to chemical poisons, bacteria, or viruses; the effects of concentration of these agents (child versus adult); and strategies for quantitative and qualitative analysis of data.

Case Presentation: Depending upon labs used, this can take up to four hours of class time, which is divided into three phases. In Phase 1 students become familiar with the setting of the camping trip and begin to imagine themselves having the experience. The phase ends with students making initial hypotheses about why a six-year-old girl gets sick. In Phase 2 students explore alternative hypotheses for the girl's sickness as her condition worsens. In Phase 3 students complete their assigned essay using their choice of hypothesis and its associated facts, symptoms, and so on. As students conduct their investigations, they connect the child's symptoms (nausea and headache) and the facts provided in the story (symptoms occurring within three days of eating apples from the orchard) with organic phosphate poisoning. In addition to satisfying their curiosity, they have also been engaged in quality science directly related to state and local curriculum objectives.

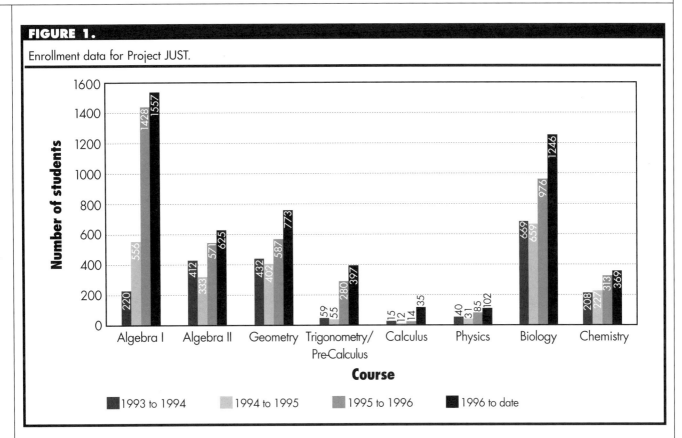

FIGURE 1. Enrollment data for Project JUST.

of community road design, geology, and human physiology. The problem-based learning concept is in its 10th year as a parallel curriculum offering for first- and second-year medical students.

Surveys of students and teachers involved indicate that problem-based learning is more effective than lectures, makes lab activities more relevant, and has everyone more excited about coming to class. Teachers report that this method provides an integrated approach that impacts self-esteem and student research skills.

FORMATIVE DATA

After a year of operating, Project JUST has seen significant increases in minority participation in upper level science and math classes. The data (Figure 1) indicate that we are moving in the right direction. The 1993–94 data reflect our benchmark number. The 1995–96 data represent our first year of operation under the National Science Foundation grant. Enrollment of minority students in upper-level math and science courses is steadily increasing. The rapid rise in Algebra I and Biology enrollments can be partially attributed to a North Carolina requirement of Algebra I and Biology for high school graduation. Enrollment increases in Geometry, Algebra II, Trigonometry/Pre-Calculus, Calculus, Physics, and Chemistry can be directly connected to recruiting initiatives.

In addition to increasing enrollment, we must also address the number of students who are successfully completing our courses. In 1995, 45 minority students successfully completed the science sequence of Biology, Chemistry, and Physics. In 1996, 88 minority students successfully completed the science sequence. The completion rate of minority students in Geometry and Trigonometry/Pre-Calculus has doubled during the same period of time.

THE NEXT STEPS

Although we are making progress, our work has just begun. There are still too many students, especially minority students, who are not getting the benefit of rigorous math and science classes. We still have upper level science and math classes with high failure rates. As we work to increase enrollment, we must make sure we do not set students up for failure. We must encourage more teachers to transform their classrooms, more parents to get actively involved in their child's education, and more community partners to align with school administrators in a focused commitment to systemically change education.

Systemic reform is necessary for the growth and survival of public schools. It is a messy and difficult job with no short cuts. The key is to put resources where essential changes can take place. In Winston-Salem and across the nation we must look within ourselves and trust that every child possesses the fundamental desire to be successful and must believe that we can make a difference. ✧

Stan Hill is the coordinator of K–12 science and the NSF Project Director in the Winston-Salem/Forsyth County Schools, P.O. Box 2513, Winston-Salem, NC 27102-2513, e-mail: sworkshill@aol.com.

Make the Curriculum

THE DOMINANT APPROACH IN PUBLIC schools to educating students is that of traditional education. This curriculum has been most effective in educating students of western European ancestry who have experienced few economic hardships. However, the United States is a nation of diverse cultures. In spite of the diversity of students' backgrounds, the traditional curriculum is heavily skewed to western European culture and ways of thinking.

Something is wrong! Articles appearing in journals and newspapers, topics of radio and television programs, and recently published academic reports point out that America's public schools are failing to properly educate students of color. The time has come for schools to move to a curriculum of inclusion in which the importance of all individuals and their contributions to American society are an integral part of everyday school experiences.

To be effective the curriculum must reflect the cultural diversity of all students. If we accept the metaphor that the United States is a tossed salad of diverse groups, then the curriculum should facilitate understanding and appreciation of the distinctiveness of individuals and their contributions to American society. Multicultural education embraces such a curriculum.

The chasm continues to widen in America between the affluent and the poor. Many people argue the condition is related to education or race. How accurate a predictor is school performance to the type of vocation a student will have as an adult? How related is school performance to the number of African Americans with good jobs in areas such as engineering, education, government, law, sports, enforcement, entertainment, and politics. American society today remains highly segregated, with African Americans most visible in the areas of sports and entertainment. In fact, it can be said that American society's infrastructure for the twentieth and twenty-first centuries remains consumed by the cancers of racism and greed. What are schools doing about these ills? How can we bring students to have hope in their

BY NAPOLEON A. BRYANT, JR.

Multicultural

Act now and make science an inclusive endeavor

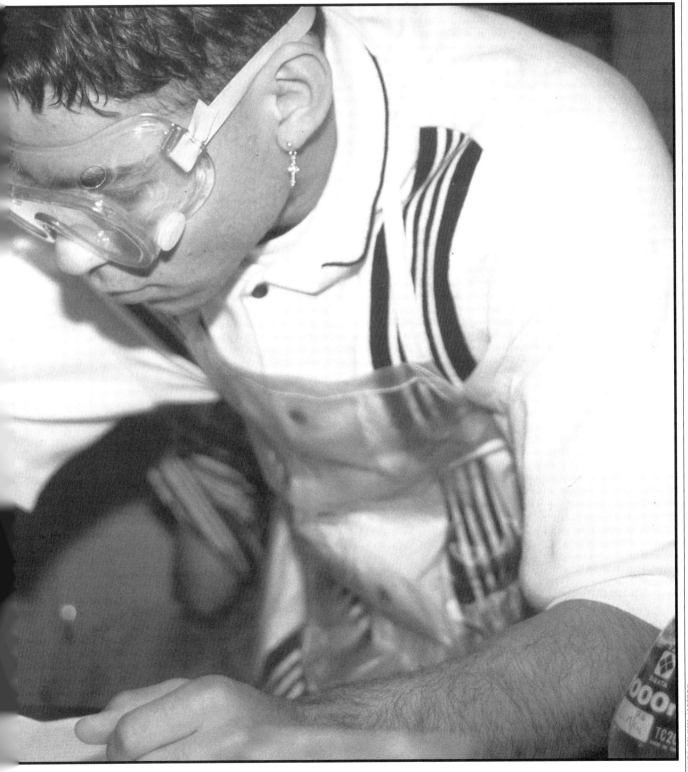

SCIENCE LEARNING FOR ALL **Celebrating Cultural Diversity**

13

abilities, to believe that one gateway to a productive life is earning a diploma from high school or, better yet, a college degree?

Many conceptions, misconceptions, attitudes, and values are based on experiences of our formative years. The results of a separate and vastly unequal society are ever present. How does the curriculum experienced by students of color prepare them to avoid the pits of functional illiteracy, unemployment, teenage and illegitimate births, dysfunctional homes and neighborhoods, social welfare, alcohol, and substance abuse? Should the curriculum deal with racism, greed, or other societal problems giving rise to the pits cited above? Should discussion of these problems, where applicable, be an integral part of the science curriculum?

The curriculum of traditional education fails to convey to students of color the interrelationship between school experience and their chances for success as adults in obtaining employment, marriage, and personal fulfillment. The goal of schools in preparing students for life, liberty, and the pursuit of happiness has eluded students of color. The curriculum of traditional education has been and remains a curriculum of exclusion. Students not of western European ancestry leave school knowing very little about their culture or contributions made to society by individuals of their culture. It is not unusual for the contributions to be treated as appendages to the curriculum or presented in an unfavorable manner.

American public schools mirror the attitudes of American society. Our public schools are at risk today. The strength of our nation is inextricably tied to the quality of our public schools and their effectiveness in educating all students. America's strength lies in its diversity and in its respect for equity among all individuals. As green plants need sunlight and nutrients for strength and growth, America needs the talents of all its citizens to retain its position of international leadership and resourcefulness. Unity is needed among America's many diverse groups. There can be unity in diversity; there can be unity with equity. Schools must emphasize this unity among students. The tool most viable to accomplish unity in diversity with equity is the school's curriculum. The curriculum must reflect the cultural diversity of the school's population; it must treat as worthwhile the contributions of all groups to American society; it must become multicultural!

Some ask, "Just what is multicultural education?" It is an educational prescription for all students. Multicultural education fosters increased self-respect, increased self-confidence, and appreciation of one's culture seen in context with other cultures. In multicultural education a greater number of similarities, rather than differences, between races is discovered with the cultural history corresponding to each student taught as an integral part of the curriculum. An overriding principle of multicultural education is the attitude: The mind of any child is too precious a commodity to waste. The minds of all children, irrespective of their economic, ethnic, or racial backgrounds must be nourished if America is to lead—to forge a new pathway that results in the improvement of the human condition at home and abroad.

Multicultural education emphasizes inclusion. That is, the contributions of all Americans are recognized and valued. Students develop increased knowledge and appreciation of their race and cultural heritage as well as that of others. Their curricular experiences foster appreciation and respect for cultural diversity with equity. Using this approach, students come to value understanding each other, working and living together, and sharing power—political, social, economic, and financial.

Multicultural education embraces cooperative learning, constructivism, questioning techniques, strategies of inquiry, creative and critical thinking, and authentic assessment of student performance. These are not new pedagogical approaches by any means. However, when used in a multicultural setting, these approaches reflect sensitivity to the cultural backgrounds of students, fostering interaction and development of mutual respect and appreciation between them.

Personal pride within students is a tremendous motivator. If we want students to smile broadly and put forth 150 percent effort, we must convince them they can master the task at hand. All students want to be successful; students of color are no exception. If we show we care about our students by holding up models from their culture and committing to assist them, students will try their utmost to achieve our expectations of them. Successful teachers engender successful students. Our expectations of students of color must be high. In multicultural education, all students are pushed to their limit because teachers value them as human beings. Given this kind of teacher and a curriculum with which all students can identify, students of color will grow to appreciate the value of an education.

Is multicultural education the panacea for what ails America's schools, its society? Can we develop a multicultural curriculum that provides students of diverse cultures with experiences that enable them as adults to work, worship, socialize, and recreate together? The laws, principles, and concepts of science are the same for all students. It is not the substance of science that gives students of color difficulty in mastery as much as it is the "negative force field" surrounding them. It is not uncommon for students of color to feel that science is beyond their capabilities. Often math and language skills, skills of observation, creativity and critical thinking, and a good self-concept are lacking in students of color. These skills and attitudes either receive little attention in the early years of students' school life or they are squashed by forces operating in the students' homes, public schools, communities, or associates. Most students of color have the ability to succeed in school. However, the creed for membership in cliques, gangs, or other detracting school groups declares being on the honor roll, making As and Bs, or enjoying school taboo. The picture is bleak but not hopeless. Increasingly educators throughout the nation are sensing the need to address the lack of success of students of color in our public schools.

CHANGE IS TAKING PLACE

The following examples of current policies and opinion on multicultural education illustrate change in our thinking from exclusive to inclusive. The Kentucky State Department of Education has issued guidelines on multicultural education that require all students to be "aware of and affirmed in their own cultural roots" (Hamer, 1990). Those leading the charge in Kentucky feel the education students receive should help them appreciate, value, respect, and understand persons of different backgrounds. The National Science Teachers Association in 1991 approved a position statement declaring: "Our global society consists of people from many diverse cultural backgrounds; we should appreciate the strength and beauty of cultural pluralism. . . . We are aware our welfare is ultimately dependent upon the productivity and general welfare of all people."

The curriculum and how it is presented is not the only part of the education process that needs overhauling. Teachers must learn to think multiculturally and become skilled in organizing learning experiences that reflect the new way of thinking. Pamela L. Tiedt says "multicultural education is an infusion process that has two components—one is multicultural teaching and one is multicultural content. It means focusing on the pedagogy, having a varied repertoire of strategies—questioning, cooperative learning, hands-on activities, lectures—as well as accepting that the teacher is not the sole source of education" (Tiedt and Tiedt, 1995). Iowa State University now requires all students to take a course designed to teach the contributions of minorities in various areas.

The common thread found in these positions is that to help students live in a world of diverse cultures,

Teachers must learn to think multiculturally and become skilled in organizing learning experiences that reflect the new way of thinking.

teachers' attitudes must change; we must change the way we think. Our thinking must shift from being exclusive to being inclusive. Only then will we be able to help students change the basis of their thinking and use the tools of interaction, cooperation, respect, and acceptance to build bridges nationally and globally. We must first master use of these tools ourselves.

VALUABLE OUTCOMES

The goals of multicultural education are incontestable. How can we dispute goals such as:

- intellectual development,
- preparation for vocation,
- acceptance of and respect for diversity with equity,
- appreciation for the fine arts, and
- development of respect for self and others through national and global awareness.

As teachers charged with the responsibility of helping shape the minds of students, we must do all in our power to help all students achieve their maximum potential in life.

The first step is to get to know everything there is to know about our students and their respective cultures and then diligently try to present school experiences both within and beyond that with which the students are familiar. Our students are worth the effort it takes. We must accept the challenge to embrace a curriculum of multicultural education for all students to experience. Multicultural education is more than just a dream. It is the right thing to do. ✧

Napoleon A. Bryant, Jr. is former chairperson and director, NSTA Committee on Multicultural Science Education and is professor emeritus of education, Xavier University, 3527 Skyview Lane, Cincinnati, OH 45213-2040.

NOTE

This article is adapted from a speech given at the Leadership Institute for Reform in Middle Schools, held June 29-30, 1995, at the University of Iowa.

REFERENCES

Hamer, I. 1994. Kentucky adopts statewide multicultural program. *The Multicultural Link* 1(3):1-3.

Tiedt, P.L., and I.M. Tiedt. 1995. *Multicultural Teaching: A Handbook of Activities, Information, and Resources.* Boston: Allyn and Bacon.

Inclusive Classrooms

A MULTICULTURAL LOOK AT THE
NATIONAL SCIENCE EDUCATION STANDARDS

IN RESPONSE TO CALLS FOR BETTER SCIENCE education, the National Research Council (NRC) issued the *National Science Education Standards* (NRC, 1996) for developing teaching methodologies and science curriculums, with the goal of achieving science literacy for all learners. The Standards stresses that all types of learners need to be included in science education and can benefit daily from the problem-solving processes, knowledge, and discipline found and learned in science.

In addition, the Individuals with Disabilities Education Act (IDEA) mandates that all children with disabilities be provided with free, appropriate public education in the least restrictive environment (LRE). This has created inclusive classrooms with students with different learning styles. Students in inclusive science classrooms may have attention deficit disorder (ADD), short attention spans, difficulty processing new words, and/or trouble "visualizing" abstract ideas. They may have sensory, visual, auditory, lingual, behavioral, and other difficulties and needs.

Because the Standards are devoted to the goal of science literacy for everyone, science educators will find the Standards especially useful in constructing and modifying lessons to create learning environments where all students, including those with disabilities, can succeed and thrive. To work effectively in inclusive classrooms, science teachers can employ five Standards-based strategies that address the particular needs of students with learning disabilities.

STRATEGY ONE

A multisensory approach maximizes the learning of students with language and/or behavioral challenges.
Learning is most effective when students are treated as individuals with different learning strategies, approaches, and capabilities (American Psychological Association, 1995). Science teachers have a variety of classroom resources that may be used to make lessons interesting, thus decreasing disciplinary problems associated with bored students and accommodating different learning styles.

Science is a natural subject for hands-on (kinesthetic) learning experiences that appeal to the visual and auditory senses. Physical, pictorial, and symbolic examples can be integrated for a multisensory approach to teaching students with specific sensory and language disabilities. In addition, hands-on scientific experiments that reinforce scientific concepts are appropriate for students with ADD or with behavioral disorders because such tasks attract students' attention and interest (NRC, 1996). Educators must create learning environments where the learning is student centered and the tasks are not only challenging but also interesting and relevant to the learner.

An example of such an approach is a lesson on friction with multisensory appeal. The lesson begins with students rubbing their hands together and discussing how it feels, which appeals to kinesthetic learners. Students then choose a sport or activity of interest to them and investigate the role of friction. To draw on the interests of the learners, the teacher shows sports videos, which appeal to visual learners, and a class discussion follows about when friction is wanted as well as instances when it is not (an activity that appeals to auditory learners). Students can contribute their own experiences, such as receiving bruises caused by falls from bikes or on slippery floors or icy pavement.

How the word friction is used in common speech can also be discussed. For instance, teachers can mention the expression "causing friction between friends" and compare it to the scientific usage of the word friction. The scientific and common definitions can then be written on the board. Interdisciplinary topics such as how friction affects the burning of dust particles as they enter the Earth's outer atmosphere or how friction influences car designs may also be discussed.

STRATEGY TWO

Encourage collaboration and mastering of scientific concepts among learners.
When the emphasis is on mastery of goals and collaboration rather than on individual performance, students achieve success in scientific inquiry. Compared with competitive and individualistic learning situations, working cooperatively with peers produces a much more positive learning experience as well as increases students' self-esteem and feelings of success (American Psychological Association, 1995). The opposite is also true. Learning is stifled in environments where individual competition takes precedence over learning as a community (New York State Education Department, 1996).

By encouraging collaboration, teachers make students aware of the importance of learning new skills and taking the focus away from individual competition. Beneficial gains in the classroom result when students are encouraged to master material learned. In this environment, students view mistakes as stepping stones to furthering their mastery rather than as failures (Ames and Archer, 1988).

Educators should encourage students to concentrate on how the right answer was derived, not just on the answer itself. Also, emphasis should not be placed on social comparison. This will de-emphasize individual weakness as well as provide focused metacognitive activities that are beneficial to both low and high achieving students (White and Frederiksen, 1998). When students perceive the emphasis of the class to be on self improvement, they are more willing to ask for help. Students who need the most help, such as those with learning disabilities, will benefit the most. In classes where individualism prevails, there will be more avoidance of asking for help (Ryan et al, 1998).

One place in which collaboration and concept mastery can be applied is the science laboratory. Students working in teams of three or four to carry out experiments achieve positive interpersonal relationships when they use their strengths to contribute to the final

KONSTANTINOS ALEXAKOS

product. When making time measurements for example, a student with ADD could do the writing, while a student with writing difficulties could do the timing. Such positive interdependence promotes a feeling of accomplishment and belonging among students.

STRATEGY THREE

Provide authentic assessment and expectations for each individual learner.

Teachers need to ensure that lessons are of the appropriate difficulty level, or students at the low or high ends may lose interest or become frustrated. Assessment can foster learning if it is used appropriately and if it measures what it actually claims to measure. Giving students feedback gives them an understanding of how well they are meeting expectations and encourages improvement. In addition, students become responsible for their own learning if given the opportunity to evaluate their own work and reflect on their accomplishments (NRC, 1996).

An important part of special education reform is the focus on competence rather than deficiency. While recognizing individual differences in the development of learners, an environment of high expectations and effort needs to be nurtured. Research shows that quality outcomes can be achieved through high expectations (Telzrow, 1999). When students with learning disabilities blame their failure on their disabilities, it has a negative impact on their academic progress; students who see their failures as a lack of effort on their part make the most gains. Intervention in altering thinking and increasing effort may improve achievement (Kistner et al, 1988). Therefore, recognition must be given for good effort. The class should be an environment in which all students, including students with learning disabilities, feel high self-esteem and have a positive connection.

Because of the challenges learning disabilities present to some students, teachers should be careful when assessing them. Written multiple-choice exams may not always be appropriate, so use of multiple or alternative methods of assessment of each student should be developed. For example, some students who are auditory learners may benefit from hearing a recording. In this instance, a tape recorder may be used to record test questions and/or answers. If necessary, a student aide can be employed to read the questions or record answers.

Assistive technology, such as computers equipped with touch-sensitive screens or speech synthesizers and voice recognition software should be employed when appropriate. Such technology can decrease the feelings of separation and isolation of learners with disabilities, reducing their frustration, and increasing their self-esteem.

Teachers should evaluate students' knowledge and process skills in personal and creative ways such as analyzing the application of scientific concepts to the creation of experiments and designs, evaluating their explanations of their results, and making observations of the individual learner while performing tasks. Rubrics provide a clear and concise written evaluation of the student's work and should be provided to the individual along with helpful and constructive comments. Figure 1 shows a sample lab evaluation sheet that may be used for this purpose.

STRATEGY FOUR

Students can develop scientific knowledge over time through repeated exposure to and integration of a particular scientific concept.

There is a limit to how much an individual can process at any one time. An effective method of learning concepts involves breaking information down into small chunks and then relating the material to larger units. With the material covered in manageable portions, students can easily process, internalize, and remember the concept. Students' knowledge can be expanded and reinforced over time if they can draw on past learned knowledge to gain insight into the new material (Shuell and Lee, 1976). Learning science should be an active process. Students need to learn to integrate many different ideas, draw relationships, and use them to explore and understand new phenomena. This process helps not only non-labeled students but also students with language disorders and ADD.

For example, many students find the concept of vectors abstract and difficult. It is unnecessarily hard to teach this concept by itself, rather than introducing it over a period of time in relation to other phenomena. When vectors are first introduced, students' recollection of maps could be used to give them a simple comprehension of the

---

FIGURE 1.

Sample lab work evaluation sheet.

A. Written team report (28 points)
- ____ of 4 points — Need for such experiment
- ____ of 4 points — Safety precautions
- ____ of 5 points — Collection of data
- ____ of 5 points — Calculations
- ____ of 5 points — Analysis
- ____ of 5 points — Neatness

B. Team work (24 points)
- ____ of 8 points — Safety
- ____ of 8 points — Collaboration
- ____ of 8 points — Individual participation

C. Verbal assessment (24 points)
- ____ of 12 points — Team comprehension
- ____ of 12 points — Individual comprehension

D. Individual written report (24 points)
- ____ of 12 points — Essay evaluating the particular experiment and outcome
- ____ of 12 points — Suggestions and criticisms concerning the experiment

Total: ____ of 100 points
Comments:

Note: Specifics of rubrics should be discussed before being made final so students can express their concerns and suggest changes.

idea that vectors represent both a quantity, such as distance, as well as a direction. To further engage their senses, they can be asked to stand up, walk around the room, and discuss their individual displacements from the window, from the door, from their seat, and so forth. As the lessons on motion progress, the concept of vectors can be applied to velocity, acceleration, and forces. Thus, with each topic, students are given a chance to reflect on the relationship of vectors to the new concepts and better internalize the knowledge of their use and applications.

STRATEGY FIVE

Create a learning environment that is nurturing and supportive for all students.

Positive emotions, curiosity, and expectations can increase students' interest in learning, while excessive stress has negative effects on motivation (American Psychological Association, 1995). A sense of belonging, relevance of tasks, hands-on experiences, curiosity, humor, and fun all contribute to classroom interest (Bergin, 1999). In classrooms where teachers show interest in students' needs, students are more likely to ask for help (Ryan et al, 1998). By planning appropriate lessons and making modifications in the teaching style and the physical space of the room, teachers can encourage each student to achieve success and competence in the subject. A physical environment needs to be created where stress, anxiety, and disorder are reduced or eliminated.

In addition, teachers must display and demand respect for diverse ideas and learning strategies. Participation by all students must be consciously encouraged. For instance, because of their higher rate of voluntarism in science, males may be called on more often than females (Altermatt et al, 1998); African American male students, on the other hand, are statistically more likely to become disconnected with science as they get older (Osborne, 1997). Teachers should strive to avoid both of these phenomena. Additionally, students with mild disabilities in mainstream classrooms are less accepted by their peers (Cook and Semmel, 1999) and in danger of becoming "invisible." Because these students may not be willing to volunteer answers, educators must ensure that no students, including those with learning disabilities, are forgotten.

To help all students feel included, the teacher can ask at the beginning of the term for personal inventory cards stating students' likes and dislikes, hobbies, and interests. These interests can be incorporated into lessons, establishing positive personal connections. For example, movies such as *Armageddon* or television shows such as *Star Trek* could provide a way to talk about mechanics and Newton's laws in space. All students can be drawn into such a discussion, making the class fun, as well as establishing a sense of community.

The Standards calls for all learners to achieve scientific literacy. This article contains five strategies adapted from the Standards to aid science teachers of classes that include students with learning disabilities. A science teacher must be flexible and willing to try new approaches and methods of teaching. Whether or not inclusive classrooms work depends on the experiences of all involved, especially the students. ✧

Konstantinos Alexakos is a physics teacher at F.H. LaGuardia High School, 100 Amsterdam Avenue, New York, NY 10023; e-mail: kalexakos@yahoo.com.

REFERENCES

Altermatt, E.R., J. Jonanovic, and M. Perry. 1998. Bias or responsivity? Sex and achievement-level effects on teachers' classroom questioning practices. *Journal of Educational Psychology* 90(3):516-527.

American Psychological Association. 1995. Learner-centered psychological principles: A framework for school redesign and reform. Washington, D.C. (Eric document ED 411 493).

Ames, C., and J. Archer. 1988. Achievement goals in the classroom: Students' learning strategies and motivation process. *Journal of Educational Psychology* 80(3):260-267.

Bergin, D.A. 1999. Influences on classroom interest. *Educational Psychologist* 34(2):87-98.

Cook, B.G., and M.I. Semmel. 1999. Peer acceptance of included students with disabilities as a function of severity of disability and classroom composition. *The Journal of Special Education* 33(1):50-61.

Kistner, J.A., M. Osborne, and L. LeVerrier. 1988. Causal attributions of learning-disabled children: Developmental patterns and relation to academic progress. *Journal of Educational Psychology* 80(1):82-89.

National Research Council. 1996. *National Science Education Standards*. Washington, D.C.: National Academy Press.

The New York State Education Department. 1996. *Learning Standards for Mathematics, Science, and Technology*. New York: The University of the State of New York.

Osborne, J.W. 1997. Race and academic disidentification. *Journal of Educational Psychology* 89(4):728-735.

Ryan, A.M., M.H. Gheen, and C. Midgley. 1998. Why do some students avoid asking for help? An examination of the interplay among students' academic efficacy, teachers' social-emotional role, and the classroom goal structure. *Journal of Educational Psychology* 90(3):528-535.

Shuell, T., and C. Lee. 1976. *Learning and Instruction*. Monterey, Calif.: Brooks/Cole Publishing Co.

Telzrow, C.F. 1999. IDEA Amendments of 1997: Promise or pitfall for special education reform? *Journal of School Psychology* 37(1):7-28.

White, B.Y., and J.R. Frederiksen. 1998. Inquiry, modeling, and metacognition: Making science accessible to all students. *Cognition and Instruction* 16(1):3-118.

Inclusive Reform

Including all students in the science education reform movement

IN THE UNITED STATES, TEACHERS, STUDENTS, and researchers are participating in a science education reform movement. There is great interest in the nation's science accomplishments and its students. Money has been allocated to improve science achievement, and unified standards for learning and teaching science have been formulated. However, whether all teachers, researchers, and especially students will benefit from this science education reform movement remains to be seen. Requisite changes will ensure the involvement of more K-12 students in learning quality science. These changes are related to organizational arrangements, science curricula, learning environments, and teaching styles. Nevertheless, more questions than answers about science learning and teaching for all students remain.

REFORMING PEDAGOGICAL POLICIES

In the United States, there have been many science education reform movements, the focuses of which have ranged from science content knowledge to the nature of science to the personalization of science. The theme of the current science education reform movement, "Science for All," embraces every K-12 student. However, involving all students in a reform has its dilemmas, many of which were not addressed by researchers until the latter part of this century.

MARY MONROE ATWATER AND MELODY L. BROWN

Until the 1920s, secondary science curricula comprised a variety of subjects that were offered as one-semester-long courses during a student's first two years of high school. From the 1930s to the 1950s, the aim of science teaching was the enrichment of students' lives so they could participate in democracy. Ironically, the accomplishment of this goal was denied to most students of color, European American females, and impoverished European American males.

In the 1960s, science education again became a priority in the United States after the Soviet Union launched the satellite Sputnik, leaving Americans to question their ability to compete in an increasingly technologically based world (Gabel, 1994). Million of dollars were spent to develop science curricular materials that covered the prevailing scientific theories, principles, and laws and to equip teachers with up-to-date science knowledge, skills, and facility in instructional methods based on the prevailing learning theories (Majumdar et al, 1991). Science curricula were designed for the "brightest and best" in science classrooms while the social problems and individual needs of students received little attention (Hurd, 1970). During the 1960s, some public schools were desegregating; others were fighting in courts to remain segregated (Commission on Civil Rights, 1977a, 1977b). Therefore, students and teachers of color and many European American students and teachers living in poor or isolated sections of the country were not included in the science education reform of the 1960s (Anderson, 1983).

In the late 1970s, there was a move "back to the basics," but science was not considered one of the basics; hence, funding for science education drastically declined (Hurd, 1979). Parents and other educational stakeholders believed that many high school students could not reason, were irresponsible, and did not understand democracy. Therefore, the 1970s reforms emphasized reading comprehension, writing, and arithmetic operations. In addition, issues related to the environment, urban schools, civic duties, and desegregation became the focus of education.

In the early 1980s, there was a call to revolutionize science education (Gabel, 1994) because many education stakeholders believed the United States was not as competitive as it could be in the world market. Some researchers attributed this lack of competitiveness to a scientifically illiterate populace (National Commission on Excellence in Education, 1983). This belief was based on scores showing that precollege students were not performing well on standardized science tests (Education Commission of the States, 1983) and the realization that teachers were not current on the latest thinking in science, science learning, and science teaching. To address the problem of a scientifically illiterate population, reformers believed science curricular materials needed to be both minds-on and hands-on (Hurd, 1983), schools needed to be safe for students to learn (Martinolich, 1979), and public schools should be held accountable for student learning (Culyer, 1988).

SCIENCE FOR ALL

The American Association for the Advancement of Science (AAAS) developed Project 2061, one of the first projects to initiate the present reforms in science education. Its first effort, *Science for All Americans* (AAAS, 1989), spearheaded this effort and delineated the knowledge and skills a high school graduate should possess. This effort was founded on the belief that high school graduates would be scientifically literate if they recognized that science, mathematics, and technology are interdependent human enterprises with strengths and limitations; understood key science concepts, principles, and theories; were familiar with the natural world; and used their ways of thinking and scientific knowledge for individual and social purposes. The National Research Council (NRC), an agency of the National Academy of Sciences (whose membership is composed of distinguished scholars engaged in scientific and engineering research who advise the federal government about scientific and technological matters), has recently distributed its standards for science education with a goal to "create a vision for the scientifically literate person and standards for science education that, when established, would allow the vision to become a reality" (NRC, 1994, I-1). The NRC (1995) also delineated standards for science teaching, professional development of teachers, science assessment and evaluation, science content, school science programs, and science education systems. In 1991, the National Science Foundation (NSF) implemented a policy to fund large grants focusing on systemic change because "fragmented policy systems" do not work and need to be replaced "with a coherent system of curriculum controls" (Fuhrman and Malen, 1991, 244).

Several assumptions underlie the theme "Science for All." First, *all* includes all K–12 students in U.S. schools (American Association for the Advancement of Science, 1989). Second, there is the belief that all students can learn quality science. Finally, a strong commitment is warranted to ensure that all students in K–12 science classrooms have the opportunity to learn quality science.

WHO STILL NEEDS TO JOIN THE REFORM?

Many of the students who have not become part of the current science education reform movement are poor and students of color or are students with disabilities. Others are English Speakers of Other Languages (ESOL) (Minicucci et al, 1995). Another group of students being excluded from the reform includes those with cognitive, social-personal, and intellectual disabilities. Students with disabilities are often homogeneously grouped in self-contained classrooms where they have little interaction with other students in the school and are excluded from science education reform.

To bring about change, educators must recognize that organizational arrangements affect the science curriculum. Both "formal and informal organizational features affect what and how teachers teach science" (Cuban, 1995, 6). These organizational structures are decided at the district, school, or classroom levels. Governance, formal structures (essential building blocks of a school), cultural processes (unexamined, deeply embedded norms and expectations that staff members share about their tasks of schooling students), and teaching are crucial components to consider in school organization.

One organizational structure is the age-graded school, a well-established organizational school arrangement based on the assumption that educational quality, efficiency, and equality are the result of uniformity. Age-graded school structures influence the actions of both students and teachers in science classrooms, curricula, schedules, and evaluation (Cuban, 1995). However, there have also been attempts to organize students into ungraded elementary classes and multi-age groupings. For example, one urban elementary school in a large school district in the South implemented this ungraded practice for grades K–3. A former colleague working with the school staff disclosed that the most difficult task was to convince new teachers that this practice aided students in the learning process and facilitated the development of self-esteem (Frick, 1985). Recently, ungraded organizational structures have been implemented in some K–12 school initiatives (Silula et al, 1996).

BORDER CROSSING

In addition to paying attention to organizational structures, educators should consider the home environment of students. Many students cross borders from the microcultures of their families and communities into the microcultures of schools and science classrooms. Border crossing of students into schools is discussed by Henry Giroux (1992), an internationally known critical theorist who addresses how power gets played out in social institutions such as schools; border crossing of students into science classrooms is discussed by Glen Aikenhead, a Canadian science education research scholar (1996). Some researchers believe that learning science involves students discerning a new culture. "The extent that students understand, investigate, and determine how the implicit cultural assumptions, frames of references, perspectives, and biases within [science] influence the ways in which knowledge is constructed within [scientific disciplines]" (Banks, 1981, 21) helps determine the success of students transversing science borders.

Science learning is a risky process, just like any other learning, and making mistakes in order to understand natural phenomena is typical among both students and scientists. Encouraging students to be risk takers and establishing environments in which students feel safe to make mistakes are requisite in understanding school science. "Walking into a science classroom where the lessons are not taught in your mother tongue" (Reichert, 1989, 10) or teachers' attempts to instruct students whose languages or cultural customs they do not understand "can be like walking into a dark cave from which there is no exit" (Reichert, 1989, 10). This kind of science classroom is not a safe learning environment for students. However, ESOL students can learn the same science curriculum as native English speakers (Minicucci et al, 1995) if necessary changes are made to enable these students to be a part of the prevailing science festival.

Thematic learning assists ESOL students in learning science and validates their cultural and linguistic backgrounds (Minicucci et al, 1995). They also found that when both science and bilingual teachers acted as facilitators, students learned science. In addition, when ESOL students were engaged in science inquiry and active discovery in small cooperative groups during thematic science learning, their limited English literacy was ameliorated.

In addition to ESOL students, many students with disabilities are excluded from science education reform; nevertheless, some science educators believe that science as a process should be incorporated in the curricula for students with disabilities (Cawley, 1994). Likewise, few comprehensive science curriculum programs exist for students with severe emotional, behavioral, or intellectual disabilities. Such students who are involved in a hands-on science program have been shown to score significantly higher on pre- and post-tests designed by the researchers to evaluate science content knowledge than those with only the textbook

> Students must ex *academic success* develop an their *cultural ident* develop a critic ness through whic can *chal status quo* of the social ord scientific knowledg

treatment (Mastropieri and Scruggs, 1993). Other successful science teaching approaches include tutoring, cooperative learning, mnemonic strategies, and self-monitoring strategies (Mastropieri and Scruggs, 1995). Earth science students with learning disabilities understand science principles and concepts when fewer science topics are taught in more depth with a problem-solving approach (Woodward and Noell, 1991).

However, in reality, there are students with severe disabilities who will be unable to understand most of the ideas found in *Science for All Americans*. Teacher assistance is even more necessary for these students to learn and develop on the basis of their own abilities and talents, and many of these students need the opportunity to learn and work with students of varying abilities to develop their self-worth (Ferguson, 1995).

Culturally relevant teaching is "a pedagogy of opposition [that] is committed to both collective and individual empowerment" (Ladson-Billings, 1995, 160). It is based on three propositions. First, students must experience academic success in science. Second, students must develop and maintain their cultural identities (Nelson-Barber and Estrin, 1995). Third, students must develop a critical consciousness through which they can challenge the status quo of the current social order with their scientific knowledge and skills.

Science teachers can provide opportunities for students to investigate open-ended science problems (many connected with their communities) and learn to persuade others by using scientific data. Teachers may still have to overcome cultural differences, however. In some Native American cultures, for instance, discovery learning is discouraged because the student environment is fraught with physical danger (Nelson-Barber and Estrin, 1995). Because discovery learning may not always be culturally relevant in teaching science, students must be taught that the uncovering of scientific theories, laws, and concepts is the primary goal of the scientific enterprise. Some Native Americans learn science by identifying relationships and changes, observing and evaluating in context, and using a circular view of time. In some Native American cultures, valued knowledge is that which draws one closer to the spirit of harmony. School science is usually taught in a way in which natural phenomena are analyzed out of context even though actual scientists often study natural phenomena within a context.

Community involvement is required if all students are to be included in the science education reform movement. Parents and the extended family's involvement in schools and science classrooms may be indispensable for science learning. Communication in the parents' native languages at parent-teacher conferences and transmission of written messages in the parents' native language are important strategies for inclusion.

Presently, several successful family, school, and community partnerships operate in urban schools. These partnerships have three common characteristics: they provide success for all students; they serve the whole child; and they allow the school, family, and other community organizations to share in the social, emotional, physical, and academic development (Davies, 1991).

THE MULTICULTURAL SCIENCE TEACHER

Science teachers are not likely to use any of these approaches unless they have undergone a philosophical shift (Hiller, 1995). Teachers' classroom practices are usually based on their ideas about what it means to learn and teach science. If science learning is perceived solely as memorization of facts by a few capable and gifted students, then few new groups of students will benefit from the current science reform. Science teacher programs must stress active participation, case studies, historical films, and experiential activities in order to facilitate this philosophical shift for teachers so K-12 science reform initiatives embrace all students. Successful multicultural science teachers share common commitments, high expectations for students, and connections with students, families, and the community.

More questions than answers still remain about science learning and teaching of all students. One pertinent question is where we go from here in science education in order to involve more students, teachers, and communities in the science education reform movement. "It may be easier to give something to everyone rhetorically than it is in reality" (Donmoyer, 1995, 34). "How do we decide how to spend limited financial resources or how to use precious commodities such as teacher and student time?" (Donmoyer, 1995, 31). Who will make these decisions? Will teachers and students spend so much time completing the many benchmarks in Project 2061 that there will be little time to engage in self-initiated, problem-oriented projects? The lack of staff development programs for K-12 science teachers, the need for quality leadership, and an inimical environment for change agents in schools also serve as barriers to a reform movement (Gabel, 1994; Anderson, 1995).

More research is needed on the developmental qualities and characteristics of students with disabilities so that teachers can be more informed about developing and modifying science programs (Cawley, 1994). Consequently, teacher involvement in identifying research problems and conducting science education research studies is critical. When science education research shifts its setting to classrooms, schools, and communities, the voices of many more students will be heard and explicated. ✧

Mary Monroe Atwater (e-mail: matwater@coe.uga.edu) is a professor in science education and adjunct professor in social foundations and Melody L. Brown (e-mail: mbrown@coe.uga.edu) is a doctoral student at The University of Georgia, 212 Aderhold Hall, Athens, GA 30602-7126.

NOTE

This article is based on the featured presentation made at the National Science Teachers Association Area Convention in San Antonio, Texas on December 14, 1995.

REFERENCES

Aikenhead, G.S. 1996. Science education: Border crossing into the subculture of science. *Studies in Science Education* 27:1-52.

American Association for the Advancement of Science. 1989. *Science for All Americans*. Washington, D.C.: American Association for the Advancement of Science.

Anderson, R.D. 1983. Are yesterday's goals adequate for tomorrow? *The Science Teacher* 67(2):171-176.

Banks, J.A. 1981. *Multiethnic Education: Theory and Practice*. Boston: Allyn and Bacon.

Cawley, J.F. 1994. *A Perspective on Students with Disabilities*. Washington, D.C.: American Association for the Advancement of Science.

Commission on Civil Rights. 1977a. *School Desegregation in Bogalus, Louisiana*. Washington, D.C.: Commission on Civil Rights.

Commission on Civil Rights. 1977b. *School Desegregation in Corpus Christi, Texas*. Washington, D.C.: Commission on Civil Rights.

Cuban, L. 1995. The hidden variable: How organizations influence teacher responses to secondary science curriculum reform. *Theory Into Practice* 34(1):4-11.

Culyer, R. 1988. Accountability as a partnership: Professionals, parents, and pupils. *Clearing House* 61(8):365-369.

Davies, D. 1991. Schools reaching out: Family, school, and community partnerships for student success. *Phi Delta Kappan* 72:376-380.

Donmoyer, R. 1995. The rhetoric and reality of systemic reform: A critique of the proposed National Science Education Standards. *Theory Into Practice* 34(1):30-34.

Education Commission of the States. 1983. *The Third National Mathematics Assessments: Results, Trends, and Issues*. Denver, Col.: National Assessment of Educational Progress, Education Commission of the States.

Ferguson, D.L. 1995. The real challenge of inclusion: Confessions of a 'rabid inclusionist'. *Phi Delta Kappan* 77(4):281-287.

Fuhrman, S.H., and B. Malen, eds. 1991. *The Politics of Curriculum and Testing: The 1990 Yearbook of the Politics of Association*. Bristol, Pa.: Falmer Press.

Gabel, D., ed. 1994. *Handbook of Research on Science Teaching and Learning*. New York: Macmillan Publishing Company.

Giroux, H. 1992. *Border Crossing: Cultural Workers and the Politics of Education*. New York: Routledge.

Hiller, N.A. 1995. The battle to reform science education: Notes from the trenches. *Theory into Practice* 34(1):60-65.

Hurd, P.D. 1970. *New Directions in Teaching Secondary School Science*. Chicago: Rand McNally.

Hurd, P.D. 1979. Back-to-basics: A critical juncture in biology education. *The American Biology Teacher* 40(3):181-190.

Hurd, P.D. 1983. State of precollege education in mathematics and science. *Science Education* 67(1):57-67.

Ladson-Billings, G. 1995. But that's just good teaching! The case for culturally relevant pedagogy. *Theory Into Practice* 34(3):159-165.

Majumdar, S.K., L. M. Rosenfield, P. A. Rubba, E.W. Miller, and R.F. Schmalz, eds. 1991. *Science Education in the United States: Issues, Crises, and Priorities*. Easton, Pa.: The Pennsylvania Academy of Science.

Martinolich, M. 1979. A National Perspective on Alienation, Involvement, and Victimization in Schools. Paper presented at the annual meeting of the American Psychological Association, New York.

Mastropieri, M.A., and T.E. Scruggs. 1993. Current approaches to science education: Implications for mainstream instruction of students with disabilities. *Remedial and Special Education* 4(1):15-24.

Mastropieri, M.A., and T.E. Scruggs. 1995. Teaching science to students with disabilities in general education settings: Practical and proven strategies. *Teaching Exceptional Children* 7(4):10-13.

Minicucci, C., P. Berman, B. McLaughlin, B. McLeod, B. Nelson, and B. Woodworth. 1995. School reform and student diversity. *Phi Delta Kappan* 77(1):77-80.

National Commission on Excellence in Education. 1983. *A Nation at Risk: The Imperative for Educational Reform*. Washington, D.C.: U.S. Government Printing Office.

National Research Council. 1994. Draft of *National Science Education Standards*. Washington, D.C.: National Academy Press.

National Research Council. 1996. *National Science Education Standards*. Washington, D.C.: National Academy Press.

Nelson-Barber, S., and E.T. Estrin. 1995. Bringing Native American perspectives to mathematics and science teaching. *Theory into Practice* 34(3):174-185.

Reichert, B. 1989. What did he say? Science in the monolingual classroom. *Science Scope* 13(3):10-11.

Sikula, J., T.J. Buttery, and E. Guyton, eds. 1996. *Handbook of Research on Teacher Education*. New York: Simon and Schuster Macmillan.

Woodward, J., and J. Noell. 1991. Science instruction at the secondary level: Implications for students with learning disabilities. *Journal of Learning Disabilities* 24(5):277-284.

Creating a Culture for Success

Teachers accommodate different learning styles and boost achievement

THE LARGE NUMBER OF AT-RISK AND dropout minority students in urban schools should be a concern to all educators. In states where graduation is tied to proficiency tests, increasing numbers of minority students in urban settings are not able to leave high school with a diploma at the end of their senior year (Dunn et al, 1991). Some students are dropping out even earlier—leaving in grades nine or ten after repeatedly taking and failing proficiency tests. It is the responsibility of teachers, families, administrators, teacher educators, community leaders, and students to reverse this trend.

At-risk and minority students seem to have more problems with science and mathematics proficiency tests than with other examination categories (Dunn and Griggs, 1994). To address this problem, math and science proficiency labs coupled with tutoring sessions are becoming common in many urban schools where high percentages of students do not meet proficiency standards. Many of these remediation attempts are having some success but not enough to make a significant difference in the numbers of students who fail to make it over the bar of proficiency level achievement (Dunn, 1997)—a problem that may worsen as legislators consider raising that bar even higher.

It is unlikely that student performance on proficiency tests will improve until educators examine instructional practices. National, state, and local goals and standards acknowledge that students learn in different ways (Trowbridge and Bybee, 1996), but do teachers apply this knowledge in the classroom? To find out, our team studied instructional strategies in a number of Central Ohio schools that provide additional assistance to at-risk minority students. We found educators working with students on a one-to-one basis or in small group labs but continuing to use their

normal instructional strategies. These efforts did not result in significant changes in scores because attempting to remediate students using strategies that are mismatched with their processing strengths does not maximize student learning potential. In other words, when teachers try to teach using the same methods, only more intensely by involving smaller numbers of students and longer sessions, they are not seeing results.

Research supports the belief that most students can learn all school subjects, but each student concentrates,

BARBARA S. THOMSON, MARY BETH CARNATE, RICHARD L. FROST, EUGENIE W. MAXWELL, AND TAMARA GARCIA-BARBOSA

processes, absorbs, and remembers new and difficult information in a unique way. How individuals learn best is defined as their learning style (Dunn and Griggs, 1995). In schools that have demonstrated increased achievement gains among underachieving students, instruction has been matched to these students' learning preferences (Andrews, 1991; Quinn, 1994; Stone, 1992). Backed by the almost three decades of learning style research that is behind the Dunn and Dunn Learning Style Inventory, our team decided to incorporate learning styles into a multifaceted action plan to help at-risk students meet proficiency standards (Dunn, 1997). More than 100 dissertations have documented increased student achievement when learning styles are identified and accommodated as students process new information.

PILOTS PROFICIENCY PROJECT

We theorized that a team of educators would need to network their individual strengths to make a difference in the lives of students who were at risk. We called the project the PILOTS (Providing Independent Learners Opportunities and Tools for Success) Proficiency Project. The entire team was charged with working collaboratively to integrate personal learning strengths, skills, and counseling to promote a culture for success. Our team included classroom teachers from Columbus public schools, teacher educators from The Ohio State University, a clinical psychologist from Independence High School, and high school seniors attending or living near Independence High School who had not passed the Ohio Proficiency Test after multiple attempts.

Three objectives were defined for the PILOTS project: identification and implementation of individual learning styles for each student, implementation of skill development strategies based on past proficiency test performance, and the creation of a culture for success and competence using personal strengths of learners.

To meet these objectives, we implemented the following plan:

- We summarized relevant learning styles research;
- Our clinical psychologist acquired test data for students who volunteered to participate;
- We discussed the culture patterns that should be developed;
- Based on multicultural and learning research, we made instructional and management guidelines; and
- All students were tested using the Dunn and Dunn Learning Style Inventory (LSI), which provides learner profiles and study skills linked to LSI strengths.

LEARNING STYLES

The students in our group were minority students who had taken the Ohio Proficiency Examination an average of six times without passing. Of the students, 86 percent were African American, 4 percent were Asian, and 10 percent were Appalachian. Our team committed to 24 hours for planned experiences with students and 12 hours for collaboration and preparation.

Each student was administered the Learning Style Inventory (Dunn et al, 1989), which has a proven record of validity and reliability (DeBello, 1990). The LSI provided learner processing strength information to the team. Each student was provided with an LSI personal profile, a computerized study skill printout of their personal processing strengths with study strategies, and individual implementation conferences. During the conferences, each student was instructed on how to use personal learning power and strategies to prepare for the proficiency exam.

Skill data obtained by our team psychologist were provided to each educator and student so we

could focus on content areas identified as weak on previous proficiency tests. These content areas for each student were explored using a variety of computerized skill building programs and reinforced by studying through their individual LSI strengths. The team provided students with a strand of knowledge skills that helped them focus on key areas they had failed to master previously. In this way, team members and students became partners in promoting understanding, accuracy, and information retrieval.

Most students were tactile and kinesthetic learners with limited auditory memory strengths. Because many students were tactilely capable (using manipulatives with their hands) or kinesthetically talented (experiencing whole body activities such as pacing while memorizing or using floor games), we emphasized studying through these personal strengths. We helped all tactile preference students, as well as those with tactile learning as a second preference, make and use flip chutes (see photos on pages 25 and 29) to reinforce new and difficult information (Dunn and Dunn, 1993).

> Students and educators worked together to meet the needs of the entire team, and the collaboration and cooperation among everyone were remarkable.

Our team psychologist also conducted individual counseling sessions with every student, and small group sessions were conducted for students with similar difficulties. In these sessions we incorporated problem-solving strategies for success. Various sessions were also conducted by a cluster of rotating team members using a collaborative learning framework.

A SUPPORTIVE CULTURE

It is critical for educators to look at their own practices when working with students who are underachieving, diverse learners. It is also crucial for students to become working partners in this process and learn how to teach themselves by matching their learning style strengths with appropriate study skill strategies. Because students are generally taught using only a few instructional strategies, underachieving students often believe they are not competent. Students who are asked to process new and difficult information day after day, year after year using non-preference styles create a whole culture of failure and low self esteem (Brunner and Majewski, 1990). We aimed to use counseling, teamwork, and favored learning styles to produce a culture for success.

Underachieving students who participated in this project came from similar neighborhood cultures. Because they shared a multi-generational social network outside the school setting, the typical interaction pattern with individual teachers at school was completely different than students' home culture (Dunn and Griggs, 1995).

To make the school culture more like home culture for these students, we had them work with a support team that included interactions with clusters of educators. Sometimes these interactions were on a one-to-one basis; at other times we had a group of 3-5 educators and 10-12 students sharing strategies. These interactions and collective support were similar to culture patterns in students' neighborhoods, which we hoped would make students feel secure and help to build a culture for competence. Students frequently shared with us that they felt very comfortable and enjoyed this type of partnership.

RESEARCH SUMMARY

Our team reviewed the learning styles research to identify strategies to use with our population of learners. We also analyzed student LSI profiles to determine similarities of our students with the research findings we discussed. Our investigations indicated that we had greater than average numbers of adolescents with the following patterns:

- Field dependent students rather than field independent. Global learners are field dependent; consequently, they need to process information by being taught through major concepts. These learners need to understand the "big picture" before they start to integrate the smaller pieces of information. We had many global processors (field dependent) but also several analytic (field independent) students who needed to process information step-by-step in a sequential pattern. Because most science and mathematics materials and courses of study are developed for the field independent learner, it is much easier for analytic students to make progress as they learn in small sequential steps. Global students, by contrast, need to be introduced to difficult materials through a major concept and then concentrate on the details. Global processors also remember new and difficult information through humorous stories, anecdotes linked to the concept, and pictures. Both global and analytic students can be achievers when they learn through their strengths (Dunn and Griggs, 1995).

- A tendency to rank low in the cognitive skill areas of analytic sequential processing (Keefe and Monk, 1986). As stated above, the LSI profiles indicated that the majority of our students needed global, conceptual processing activities.

- Preferences for bright light, quiet, warm temperatures, and informal design. All students like to feel physically comfortable when learning new information and all have different preferences (Dunn and Griggs, 1995). Our students were no exception—their LSI profiles indicated a majority of students wanted a bright, quiet, warm, and informal environment. The students migrated toward the areas that naturally accommodated their preferences. The typical school chairs in the media lab were the last to be used, although most students liked the padded chairs in the computer lab. Although most schools provide plastic or wooden chairs, research indicates that students learn better when they are allowed to use casual furniture, padded seats, or carpeted floors (Hodges, 1985; Shea, 1983; Dunn and Griggs, 1995).

- Highly parent- and teacher-oriented peer learners who reject variety and appreciate having authority figures present while learning. Underachieving students seem to have many sociological preferences and are usually good candidates for cooperative learning opportunities. However, when these students experience success, they do not want to have a variety of new alternative learning strategies imposed.

- Afternoon and evening peak hours. During each 24-hour period, students experience one or more energy highs. The time of day when these occur varies—this can be likened to the ideas of being an "early bird" or "night owl." Within our student population we had a huge number of late afternoon and evening processors whose LSI chronobiological scores indicated this was the best time of day for them to learn new and difficult information.

- A preference for intake and mobility. It is important for students with high mobility needs to move from site to site as they focus on their work. We not only had many high-mobility learners but also discovered that many of our students required intake while learning. Research shows that some people need to have food in order to be productive while working.

- Visual-kinesthetic learners who reject the auditory modality but prefer some tactile resources (Dunn and Griggs, 1995). Most of our PILOTS students had strong visual, tactile, and kinesthetic strengths, but few had auditory capabilities for processing new information. Most learners who are underachievers are not auditory processors, although much of the information given in a traditional classroom is verbal.

Because most students were evening learners, we offered the PILOTS Proficiency Project from 6:00 to 9:00 P.M. Monday through Thursday for three weeks prior to the proficiency test. Our team met and worked collaboratively from 6:00 to 7:00, during which time we prepared for the session, clarified activities, and made modifications based on the previous evening's work. Students worked with the team from 7:00 to 9:00. We scheduled the media center, the computer lab, and several classrooms for the sessions, and extra optional review sessions were held prior to the day of the test to accommodate last-minute requests for help.

We found that most students were indeed at their peak energy level for these evening sessions, while many of the teachers indicated that they were not. "Early bird" teachers had some challenges working out of their energy highs with "night owl" students, but they remained motivated and focused.

Because most students in the PILOTS Project were global processors with styles related to sound, light, design, intake, and mobility, we wanted to accommodate their preferences as much as possible. Students taught with interventions that are congruent with their cognitive processing style experience a significant increase in academic achievement (Van Wynen, 1997).

In all rooms except the computer lab (where padded chairs were already available), we provided the informal seating options that are important for successful learning (Hodges, 1985). Students were required to sign in but were allowed to be as mobile as necessary within the entire wing of the building we were using. We also experimented with food for high-intake students. Snack food, fast food, sweet food, and salty food—we had it all. We encouraged students whose profiles indicated they were high-intake learners to bring their own snacks, and they were allowed to eat most of the evening.

However, clean hands and absolutely no food were rules for both the computer and science labs, and we had no difficulties getting students to comply with these rules. All students seemed willing to ensure food would not become a problem. Several classroom teachers were impressed with the behavior and reliability of some students who did not seem to be able to function this way during regular school hours.

Because most high school underachievers rank lowest on sequential processing (Keefe and Monk, 1986), which is highly correlated to field independence, we decided to use a field dependent and global processing model. Understanding the concept and then working on the details was our focus. Visual, tactile, and kinesthetic opportunities, including the flip chutes, were provided as students worked on their challenging content areas.

> We created a culture of success by helping students understand their strengths and providing ways for them to teach themselves.

Individual counseling, small group content focus sessions, computerized skill building, and learning style conferences were other ongoing activities. Students and educators worked together to meet the needs of the entire team, and the collaboration and cooperation among everyone were remarkable. Most participants were exhausted after the evening sessions.

When students repeated the test, we were all enthusiastic but also apprehensive. We tried to prepare them for the environmental, psychological, and biological shift to the actual testing experience. The students had to take a test they had repeatedly failed in a strange location, in the early morning hours with people they did not know, all without the opportunity to eat or be mobile. Although students had been able to learn using their strengths during the PILOTS Proficiency Program, they had to take the test using some of their weakest processing areas.

Waiting for the test results was perhaps the most difficult part of the project. When the principal finally received the results, we were stunned. Overall, 75 percent of the students passed, and 21 percent had a net gain of 27–45 points. (A score of 200 was needed in order to pass the test.) We were amazed that so many students could increase their

scores so substantially in such a short time. Although we were disappointed that 25 percent of the students did not pass the exam, they are being provided with extra assistance prior to the next round of testing as we replicate our model and refine some of our strategies, and we are hopeful they will pass in the future.

These proficiency results were phenomenal for underachievers who had previously failed the test. We were told by some experts that helping these students was an impossible task, but we were all willing to take that chance, including the students and families who believed in our commitment to a community of learning. Through the PILOTS Project we tried to create a culture of success by helping students understand their strengths and providing ways for them to teach themselves. ✧

Barbara Thomson is an associate professor at The Ohio State University, College of Education, School of Teaching and Learning Mathematics, Science, and Technology (e-mail: bjsthomson@aol.com); Mary Beth Carnate (e-mail: carnate.1@osu.edu) is the development director and Richard Frost is a clinical psychologist, both at Independence High School, 5175 Refugee Road, Columbus, OH 43232; Eugenie Maxwell is a science teacher at Hilliard Darby High School, 5410 Victoria Park Court, Columbus, OH 43235 (e-mail: maxwell.44@osu.edu); and Tamara Garcia-Barbosa is a doctoral candidate at The Ohio State University, 980 King Avenue 4-6, Columbus, OH 43212 (e-mail: TJBarbosa@unforgettable.com).

REFERENCES

Andrews, R.H. 1990. The development of a learning styles program in a low socioeconomic, underachieving North Carolina elementary school. *Journal of Reading, Writing, and Learning Disabilities International* 6(3):307–314.

Brunner, C., and W. Majewski. 1990. Mildly handicapped students can succeed with learning styles. *Educational Leadership* 48(2):21–23.

DeBello, T. 1990. Comparison of eleven major learning styles models: Variables, appropriate populations, validity of instrumentation and the research behind them. *Journal of Reading, Writing, and Learning Disabilities International* 6(3):203–222.

Dunn, R. 1997. The goals and track record of multicultural education. *Educational Leadership* 54(3):74–77.

Dunn, R., and K. Dunn. 1993. *Teaching Secondary Students Through Their Individual Learning Styles*. Boston: Allyn and Bacon.

Dunn, R., K. Dunn, and G.E. Price. 1989. *Learning Style Inventory*. Lawrence, Kans.: Price Systems.

Dunn, R., and S. Griggs. 1995. *Multiculturalism and Learning Style*. Westport, Conn.: Praeger.

Dunn, R., T. Shea, W. Evans, and H. MacMurren. 1991. Learning style and equal protection: The next frontier. *The Clearing House* 65(2):120–122.

Hodges, H. 1985. An analysis of the relationships among preferences for a formal/informal design, one element of learning style, academic achievement, and attitudes. Doctoral dissertations, St. John's University. *Dissertations Abstracts International* 53, 2791A.

Keefe, J., and J. Monk. 1986. *Learning Style Profile Examiner's Manual*. Reston, Va.: National Association of Secondary School Principals.

Quinn, R. 1994. The New York state compact for learning and learning styles. *The Learning Styles Network Newsletter* 15(1):1–2.

Stone, P. 1992. How we turned around a problem school. *The Principal* 71(2):34–36.

Trowbridge, L., and R. Bybee. 1996. *Teaching Secondary School Science*. Englewood Cliffs, N.J.: Prentice Hall.

Capitalizing on DIVERSITY

Strategies for customizing your curriculum to meet the needs of all students

REGARDLESS OF THEIR CULTURAL identity, students want to learn, produce more, and increase their level of achievement (O'Brien, 1989). The challenge that science teachers face is enabling students of all ethnic and cultural backgrounds to accomplish these goals. Teachers must employ cultural knowledge, cultural sensitivity, and interpersonal skills when working with students; explore and develop expanded or new ways of teaching; and provide opportunities for students to maximize their learning potential.

As the population of the United States rapidly becomes more racially and ethnically diverse, students from various cultural backgrounds bring unique learning preferences and styles to the science learning environment. Cultural differences due to ethnicity, language variations, social class, and individual differences may influence these preferences and styles, which also will change as students age and gain experiences. There are also intraethnic and interethnic learning differences due to these cultural factors. Demographers predict that persons of color will constitute one-third of the U.S. population by the turn of the century (CMPE, 1988) and that American classrooms will begin to reflect a new majority—students of color, low-income students, and students of non-English-speaking backgrounds (Gomez and Smith, 1991).

Teachers are continually challenged to meet the needs of all students, but understanding students' individual learning preferences is especially challenging because these preferences develop from early *and* continuing socialization patterns. Because learning preferences and styles have been identified as important variables in the scholastic success (or failure) of ethnic minorities in the United States, determining how these students process information in a science classroom can provide valuable information on which to base responsive instruction and educational experiences.

Recognizing that cultural differences influence students' learning experiences, the Learning-How-To-Learn coordinator and teachers in the program Science Bound were interested in knowing more about ethnic minority students and their learning preferences. Science Bound is an extension of (1) a two-year pilot project funded by the Iowa State University College Bound program and Ames Laboratory of the U.S. Department of Energy and (2) a three-year, $400,000 mentoring grant through the National Science Foundation. It is a partnership between Iowa State University, the Des Moines Community School District, and technology-based businesses working together to increase the participation of ethnic minority students in science and technology and encourage them to prepare for careers as scientists, engineers, and technologists. University professors and district teachers plan and implement science- and mathematics-based curricula, including various programs and activities. Mentors help students see the applications of math and science as they relate to career opportunities, thus nurturing their science and technology potentials.

BY LENOLA ALLEN-SOMMERVILLE

"There is more than one way to learn science, and every student has a unique learning preference, style, and approach."

Learning-How-To-Learn, a component of Science Bound, includes, but is not limited to, goal-setting activities, becoming aware of preferences for studying and ways of most easily learning and remembering information, and finding out how to motivate oneself. Students who remain in the program for five years (grades 8–12), are accepted to Iowa State University in good standing, and pursue a technology-based field of study receive a tuition scholarship from the university.

A priority of Science Bound is to work with students to assess how and when they process information and to determine their instructional preferences. One resource for diagnosing cognitive preferences and style is the cognitive domain of the National Association of Secondary School Principals' Learning Style Profile (Foriska, 1992). The data supplied by the Learning Style Profile (LSP) allow teachers and others who work in teaching situations to diagnose students' strengths and weaknesses, ascertain students' learning preferences, and develop or modify instructional design and strategies to better match students' learning styles.

During the 1992 school year, 95 Science Bound students (American Indians, African Americans, Hispanic Americans, and Asian Americans) were administered the LSP. The LSP was hand scored in a three-stage process: scoring individual items, generating subscale raw scores, and determining subscale standardized scores. The score on a given subscale is the standard score for that subscale. The process of standardization was based on national data that were divided into categories of weak, low, or high; average or neutral; and strong. Reviewing these ranges provided the most useful basis for discussion of student and class results.

Analysis of the LSP showed that the majority of students were average to strong in willingness to work at a task until completion and in analytic, spatial, discrimination, and categorization skills. They had deficits in detecting and remembering subtle changes in information (memory skill) and grasping visuospatial relationships (simultaneous processing skill). Most students preferred hands-on or visual information to auditory. Fewer indicated less willingness to express opinions or speak out. Nearly half the students preferred working in small groups or pairs. As expected, members of the four ethnic groups also showed different learning preferences, with significant differences found between African Americans and Asian Americans. African Americans preferred categorization and verbal risk styles, whereas Asian Americans preferred spatial styles.

The Learning-How-To-Learn coordinator made the results of the LSP available to teachers, students, and parents and discussed how the information could be used to improve the student's learning outcome (at school and home). The student profiles showed that Science Bound students have the potential and capacity for science development and achievement. They can learn science. Students with strong cognitive skills (analytic, spatial, and categorization) are more ready for challenging instruction and are more capable of working at or beyond their grade level, whereas students with deficiencies in those areas benefit from focused problem-solving activities. Similarly, students with strong visual responses are likely to be less effective learners if instruction is strictly verbal (auditory). These students require assistance and practice to better appreciate and accept auditory information. Conducting these kinds of cooperative projects encourages cross-cultural communication and opportunities for academic peer influence.

EIGHT SUCCESSFUL FIELD-TESTED STRATEGIES

Identifying student learning preferences is one method for ensuring that students in multicultural settings learn science. Science teachers cannot expect to use the same traditional environments, instructional practices, and methods that were used in the past. We can no longer depend on old paradigms because they do not tap the academic potential of all students. We must become innovators in the quest to meet students' learning needs (Foriska, 1992). A concerned and creative science teacher may replicate or modify the following strategies that have been successful for Science Bound (SB) teachers, mentors, and students:

1 **Assume that students can learn.** Avoid stereotypes that limit student success in selected subjects. SB provides opportunities for students to visit a university campus and attend classes taught by professors, observe scientists in laboratories, work with mentors in the private and public sector, interact with college students, conduct research projects, and learn ways to reduce barriers to academic achievement.

2 **Use exciting and challenging hands-on activities.** Experienced teachers have indicated that instructional methodology can no longer be limited to using the science textbook supplemented with a few hands-on activities. Textbook science can turn off large numbers of ethnic minority students if it is the only way science is presented to them. Past SB students interned with the state Department of Natural Resources in the areas of fisheries, wildlife, and law enforcement. They took samples of fish to determine a lake's stocking needs, captured and banded geese and wood ducks to establish mortality and survival rates, talked with landowners about cropping for wildlife enhancement, and worked with the boating patrol to guarantee code compliance and check fishing licenses.

3 **Talk to students about their learning styles.** Encourage students to take more control of their learning and capitalize on opportunities for success. Through

SB staff, students have learned interesting things about themselves, such as the way they learn and study. After discussing their LSP profiles, students made comments such as, "I have strong spatial skills just like a scientist;" "This means I do better in the morning and should schedule hard classes in the morning, not my last two periods of the day;" "My analytical skills are weak;" and "I work best in small groups and hate to talk in class."

4 Develop a repertoire of content strategies and activities. Include a variety of cognitive/learning modes in your teaching style and assist students in moving from one preferred learning mode to a base of mixed preferences so they can benefit from information closely related to science. A 1992 summer algebra enhancement seminar introduced students to the basics of algebra through visual techniques and also provided a citywide support system to all SB newcomers. A later seminar explored algebra concepts with manipulatives.

5 Learn about the history and culture of the various groups. To successfully integrate multicultural content into the curriculum, SB teachers and mentors attended several sessions in which cultural information was disseminated. At one session, representatives of four ethnic minority groups in Iowa (Iowa Commission on the Status of African Americans, Iowa Commission on Latino Affairs, Iowa Governor's Advisory Board on American Indian Issues, and the Bureau of Refugee Programs) provided demographic and cultural information. This included, but was not limited to, characteristics of the population, family structure, customs, values, advocacy groups, organizations, and social-cultural concerns. The district's minority achievement program coordinator presented each participant with a report regarding the status of ethnic minority students in the district.

6 Help students see themselves as future scientists and appreciate the multicultural history of science. Students must realize that all cultures have made significant achievements in scientific fields and that a career in science is an exciting and realistic option for all people. Provide ethnic minority role models. Do not underestimate the importance of seeing someone successful who "looks like you" doing and talking about science. SB made concerted efforts to recruit ethnic minority mentors (male and female). The mentors work in hospitals, state government, research agencies, and the armed forces. Also, students have been provided lists of ethnic minorities who have made and are making important contributions to science and technology, such as Mae Carol Jemison, the first African American woman to fly in space.

7 Build opportunities for success into the curriculum and create climates conducive to learning. Avoid selling ethnic minority students short and underestimating their ability to succeed (Henson, 1975). As a result of their efforts in a rocket-launching project at the annual picnic, several SB students won scholarships to the U.S. Space Camp in Huntsville, Alabama, as participants in one of NASA's youth programs.

8 Provide diverse learning experiences. Engage students in both in-class and out-of-class learning opportunities. Field trips, such as visits to the Chicago Museum of Science and Industry, the Field Museum of Natural History, and the Shedd Aquarium, are enjoyable and educational.

THE MULTICULTURAL CLASSROOM

Science teachers are encountering an increasingly broad range of students in their classrooms. These students bring individual and cultural differences that should be acknowledged as opportunities rather than deficits. As science teachers recognize this, they will realize that there is more than one way to learn science and that every student has a unique learning preference, style, and approach. The inclusion of the eight strategies in the curriculum has been quite beneficial for SB students. Students' grades, motivation, cooperation, self-esteem, and attendance have all improved. Because these strategies met with such positive results with this multicultural group, they may also have success with students in other multicultural settings. ✧

Lenola Allen-Sommerville is instruction parent family coordinator for Science Bound and an assistant professor of curriculum and instruction at Iowa State University, N165B Lagomarcino Hall, Ames, IA 50011-3190.

REFERENCES

Banks, J.A. 1988. Ethnicity, Class, Cognitive, and Motivational Styles: Research and Teaching Implications. *Journal of Negro Education* 57(4): 452–466.

Commission on Minority Participation in Education and American Life. 1988. *One-Third of a Nation.* Washington: The American Council on Education.

Foriska, T.J. 1992. Breaking from tradition: Using learning styles to teach students how to learn. *Schools in the Middle* 2(1): 14–16.

Gomez, M.L., and R.J. Smith. 1991. Building interactive reading and writing curricula with diverse learners. *The Clearing House* 64(3): 147–151.

Henson, K.T. 1975. American schools vs. cultural pluralism. *Educational Leadership* 32: 405–408.

O'Brien, L. 1989. Learning styles: make the students aware. *NASSP Bulletin* 73(519): 86–89.

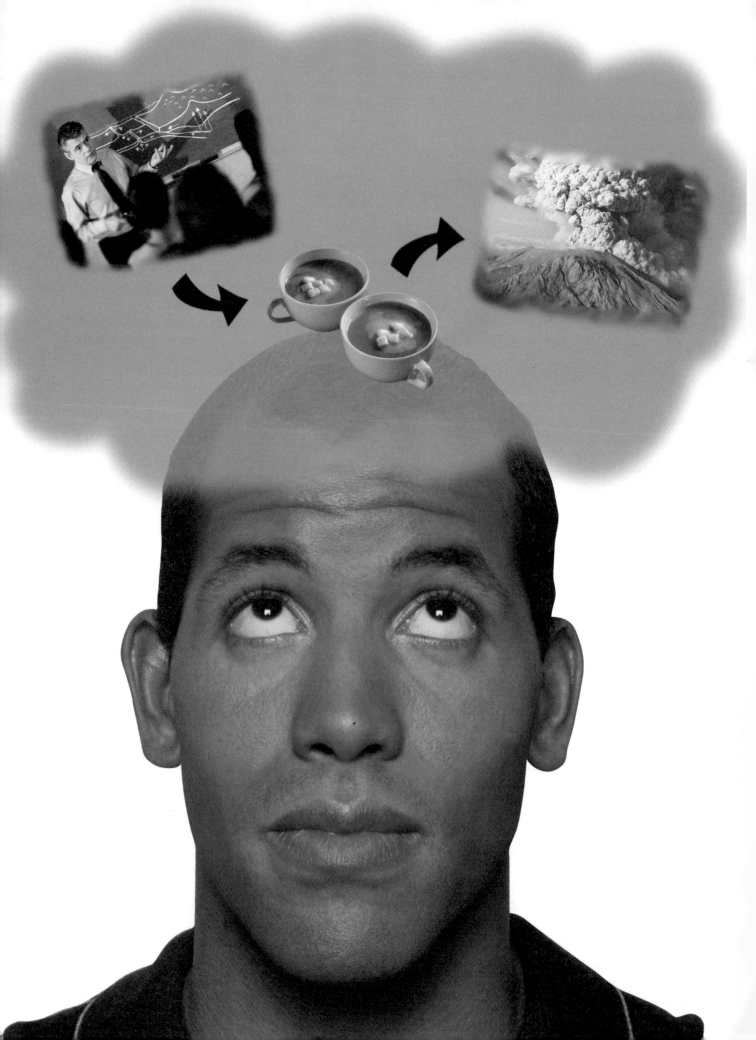

BIG Picture Science
Uncovering teaching strategies for underrepresented groups

PEOPLE FROM GROUPS THAT ARE UNDER-represented in science—females, American Indians, African Americans, and Latinos—have much to offer our educational system but often have difficulty expressing their ideas and participating in classrooms where traditional methods of instruction are used. Some experts suggest that this difficulty occurs because these groups may feel uncomfortable with the limited roles in which traditional teaching methods require students to fit (Thomson et al, 1999; Delpit, 1995). To change this situation, teachers from Jefferson Middle School in Eugene, Oregon, and Mission College in Santa Clara, California, developed the Connections Across Cultures (CaC) project to ascertain the root of the problem.

We have spent the last seven years talking to hundreds of people from these four groups—including elementary through graduate level students, teachers, community and tribal leaders, and education administrators—to find new ideas and insights about teaching science. Through this process, we discovered many innovative ways to present scientific material to our diverse student population, both males and females. For example, we found that many members of these four underrepresented groups learn better when information is presented as being part of a large framework, as being personally relevant to the learner's life, and in conjunction with a personal analogy.

The project operates out of Jefferson Middle School. Support from the National Science Foundation allowed the project to include more than 50 researchers and evaluators—K-16 teachers and community leaders in five states.

The CaC project members interviewed more than 200 people who represent the target populations of females, American Indians, African Americans, and Latinos; drew some conclusions from the responses in these interviews; and obtained feedback on the conclusions by talking with more than 150 additional people from these underrepresented groups. Almost all interviews were conducted in person and involved questions about the participants' backgrounds, education, and teachers. The interviews focused both on general education experiences and on specific experiences with and ideas about math, science, and technology. These interviews made up a critical component of the project, which also comprised extensive research from other sources, namely more than 500 references from books, magazines, and journals, classroom research, and textbook analyses.

In addition, the majority of the people who worked on the project were from underrepresented groups themselves. Of the 25 people who have worked on the project in the last three years, four were American Indians, eight were Latinos, six were African Americans, three were European Americans, and four were Asian Americans. Twenty-one of these were women.

BIG PICTURE LEARNING

One of the common themes the project members found was a need for students to personally relate to classroom material. Many teachers realize that science needs to be relevant to students and have responded by using examples that they feel would be of interest or familiar to students.

However, the research suggested ways to go an important step further. People from the four underrepresented groups wanted to bring to science what is meaningful in their own lives, but they also needed to see the big picture first.

Many of us were taught and may still teach science in a linear fashion. The focus is on course content, and we pass through the material progressively from one step to another, similar to walking down a path of stepping stones between two end points. "Stacking up facts," one on top of the other, is another image of a traditional process. This process is similar to placing one piece of paper over another until there is a high stack of papers. Both of these models depend on the ability of the students to learn, or merely accept and then remember, the previous steps as the process moves along. These teaching strategies work for some students, but many students are left behind at various points because they do not process information in a linear manner.

CHARLOTTE BEHM

For example, the majority of those interviewed by the CaC project did not appear to be linear learners. They described their learning process as one in which they must first have some sort of structure or understanding on which to hang the information and facts that come later. Without a comfortable structure or reference point at the beginning, these learners will likely get lost because subsequent information or facts have little meaning or order to them.

SCIENCE BY PERSONAL ANALOGY

For instance, students find that if they can develop a personal analogy to a science concept, this serves as a structure on which to build later knowledge. We observed the process of using personal analogies as introductory structures or reference points last summer in a workshop we gave to math, science, technology, and engineering faculty from colleges and universities throughout Oregon. In a session of about 30 people, we asked faculty to describe some of their curricula.

One man and woman team-taught a geology lab. They described teaching about a process in which tectonic plates push against each other and eventually cause an eruption on the Earth's surface, and they explained some lab experiments they used to illustrate this concept.

One woman in the group was not sure she understood the science they were discussing. She asked them if it was similar to two people who lived together for a long time, and then one day, one person pushed the other too much and they had a big fight (blow-up) because of bottled-up tension.

Most people in the group liked and appreciated the analogy. The woman looked proud of herself and was ready to hear more about the experiments. However, the geology teacher who presented the scientific material stopped the process, concerned that the analogy would degrade the quality of science being discussed because it did not exactly match the geological process. Other people in the room objected, saying that using analogies was an important process for them because it was helping them understand the concepts. Then, other teachers voluntarily presented their personal analogies, to check if they understood the phenomena.

The fourth analogy offered met the approval of the geology professor. A woman suggested that the geological process was like a cup of hot chocolate with a marshmallow on top. When she poured in hot water, the marshmallow bolted up out of the liquid.

Other comments from our study confirm this connection between analogies and understanding. According to Robin Brooke, educational director of Southern Oregon Women's Access to Credit (SOWAC) in Medford, Oregon: "It's important that students get to tell their story. Some students don't actually 'get it,' until they've told their part—a reality check for them against their life and what they are hearing. At SOWAC, the teacher presents a concept. Then students, one by one, check out their understanding of the concept, relating it to something in their own lives and checking out if they made the right assumptions" (Brooke, 1997).

Another interviewee, auto mechanic Jenny Potter in Eugene, Oregon, used an analogy as a reference point for understanding a work-related task: "One day my boss was explaining about adjusting the timing—you must look at the setup from the right angle to get it right. I asked him if it was like knowing what angle to view a shot from in playing pool, as I'm good at that sport" (Potter, 1998).

Adding a new first step, such as the formulation of an analogy, to the education process can make a tremendous difference for nonlinear, big-picture thinkers. These students will be able to develop a structure and an understanding (which may be emotional or intuitive as well as logical) on which they can organize and place the details that come later.

Introductory analogies or summaries present the overall concept, the cause and result, and the interconnections. Analogies do not have to perfectly match all the details of the scientific concept. If a student presents her or his own analogy, a teacher can validate the student's contribution by commenting on the similarities between the student's personal analogy and the scientific principles. The discussion that follows can address both the similarities and differences between the scientific phenomena and the students' analogies—a discussion that can be enlightening and fun for both students and teachers.

Using analogies can engage nonlinear learners, and linear learners will typically just downplay the big picture section and begin with their traditional ordering of information when it is presented.

CREATING CONNECTIONS

We found that many students have difficulty understanding science not because they are innately unable to understand the material, but rather because they are not able to organize and process the material in ways that

> *Students find that if they can develop a **personal analogy** to a **science concept**, this serves as a **structure** on which to build **later knowledge**.*

make sense to them. Many students who do not "get it" when studying math or science are looking for something that is not usually there in traditional math or science teaching. These students are searching for some kind of personal connection with the material that will make them feel comfortable with math or science and feel confident with the content.

While personal analogies can often create this connection, sometimes other types of connections are helpful to students. For example, a student came to the CaC office with a question about vectors from her high school physical science class. She said that she could do the steps her teacher told her: represent forces by arrows, place them head to tail, compute the angles, and find the new forces. However, she did not understand what was really happening.

Without making any verbal explanation, a CaC teacher simply asked her to stand up and put both of her arms in front of her, with one arm pointing up about 45 degrees, and the other one pointing 45 degrees downward. The teacher pulled each arm along its axis, and as a result, the student moved forward in a resulting direction that was different than each individual arm.

The girl jumped up and down, exclaiming, "I got it, I got it!" Just to make sure she understood, the teacher quickly drew the corresponding vector diagram, supplying the facts and details to fill in the big picture. When they later went out to her car, the student ran up to her mother and positioned and pulled her mother's arms so that her mother could "get it," too.

Science teachers can make science real and understandable to our diverse student population by using many of these big picture connections. Strategies for doing this include:

- *Student-generated connections.* Teachers ask students for analogies and then discuss the differences and similarities of their analogies and scientific phenomena.

- *Teacher-generated analogies.* Teachers present similar concepts/analogies in other parts of nature, or real life. The concepts do not have to be exact matches of every detail, as they provide excellent opportunities for comparing and contrasting scientific phenomena with the analogies. For example, when beginning a discussion of density, a teacher could ask students to compare the density of people in the auditorium during study hall with the density of people during an all-school meeting. The students will likely answer that the density is higher at an all-school meeting. This exercise allows students to use the word density and become comfortable and confident with it, whereas before they may have become scared or alienated by scientific jargon and equations.

Or, teachers can present topics that demonstrate how phenomena feel when they happen, or find other ways to demonstrate the content that does not involve words or equations (for example, vector forces can be illustrated by pulling arms or other types of charades).

- *Teacher-generated summaries.* Teachers write out, in 100 words or less, a summary of the essence of the information that will be covered in a class. This overview is presented to the students at the beginning of the class.

By using these and other nonlinear methods of instruction, teachers can engage a higher percentage of their students, including those from underrepresented groups; help them understand science; and allow them to contribute fully to the science classroom. ✧

Charlotte Behm is the director of the Connections Across Cultures Project, Jefferson Middle School, 1650 W. 22nd Avenue, Eugene, OR 97405; e-mail: behm@eug4ja.lane.edu.

REFERENCES

Brooke, R. Interview with author. Medford, Ore., May 1997.

Delpit, L. 1995. *Other People's Children: Cultural Conflict in the Classroom.* New York: New Press.

Potter, J. Interview with author. Eugene, Ore., February 1998.

Thomson, B.S., M.B. Carnate, R.L. Frost, E.W. Maxwell, and T. Garcia-Barbosa. 1999. Creating a culture for success. *The Science Teacher* 66(3):23.

Multicultural Teaching Tips

Practical suggestions for incorporating the diverse history of science into the classroom

MULTICULTURAL EDUCATION IS a meaningful opportunity for students to achieve their maximum potential. It is clear that in any society or classroom cultural differences exist among all students. In reality, these differences are a nation's strongest assets. Multicultural education recognizes all students' individuality and welcomes the contributions of all groups in a society. A modest attempt to explain the benefits of multicultural education and some practical suggestions for implementing it follow.

Diversity brings a wealth of resources to the classroom in the form of opportunities for cross-cultural interactions among students. As teachers, we must promote mutual respect and trust among students while we help students counteract bias and enable them to distinguish myth from reality. For instance, we must be careful not to lump together all Asian Americans or Native Americans in one group. There are distinct variations and unique cultural features within these populations as in all populations. Therefore, each student should feel proud of his or her cultural heritage and should know that his or her ethnic group is important.

The NSTA Position Statement on Multicultural Science Education encourages teachers to present top-quality science instruction that helps all students experience success. Moreover, the statement stresses the selection of curriculum materials and instructional strategies that reflect diversity. Finally, the position statement recommends that students from diverse cultures should learn about career opportunities in science, engineering, and technology. In a nutshell, educators must ensure that the contributions of minority cultures are not overlooked in the process of instruction.

Teachers must encourage the majority culture to recognize that the contributions of minority cultures are essential for the well-being of a democratic society. Learning about different cultures should start at a very young age. Teachers should stress to students that recognizing and accepting cultural differences in a classroom or in a society are essential for general harmony and peaceful coexistence.

To promote science literacy, a teacher must be prepared to reach out to all students. Teachers need to pay attention to cultural differences in their classroom and use these differences effectively in the teaching process. Students should learn that science was not solely an invention of Europeans. Science was developed throughout the ages with the contributions of many people around the world. For instance, in ancient Egypt and Mesopotamia, copper was extracted from ores, glass was made, fabrics were dyed with natural colors, and distillation was used to produce perfumes.

In the past 100 years, scientists from all over the world have helped advance scientific knowledge. Teachers should discuss the contributions of scientists from diverse backgrounds as well as from both genders. It is well established that students need role models, and these scientists can serve as role models. Figure 1 lists some scientists from traditionally underrepresented groups.

We live in an interdependent world that is com-

BY S. WALI ABDI

monly referred to as a global village. There are many problems in this global village that need the attention of scientists; no single group can solve all of these problems. Students who may aspire to become scientists should be alerted to world problems. For example, students should discuss the issues of pollution, famine, overpopulation, and disease from a global perspective. Solutions to these problems should be discussed in terms of global impacts, resources, and limitations.

In American schools, cultural diversity has become a reality. Actually, it was inevitable because of the continuous influx of immigrants from abroad and the mobility of citizens within the country. Many challenges come with multiculturalism, such as multiple languages and gender issues. Teachers need to be aware that language barriers may cause some students to require more time on in-class tasks, assignments, and laboratory activities. Teachers must give equal attention and opportunity to male and female students. According to one researcher, "girls are less likely than boys to participate in class discussions" (Banks, 1993, 3).

Teachers' attitudes and interests play important roles in using diversity among student populations to its fullest potential, promoting the notion that humanity has benefited from diversity. To meet the needs of all students, maximize learning, and promote a multicultural atmosphere in the classroom, science teachers should consider the suggestions that follow.

FIGURE 1.
Scientists from groups traditionally underrepresented in science.

African American Scientists
- **Benjamin Banneker**—Mathematician and astronomer.
- **George Washington Carver**—Chemurgist; known for his work with peanuts, improved crop production.
- **Benjamin Carson**—Pediatric neurosurgeon at Johns Hopkins University.
- **Charles Drew**—Researcher in blood preservation; work led to founding of blood banks.
- **Shirley Jackson**—Physicist; first African American woman to earn doctorate degree from the Massachusetts Institute of Technology.
- **Percy Julian**—Chemist; developed cortisone and a drug to treat glaucoma.
- **Ernest Just**—Biologist; completed major work in the study of cells and fertilization.
- **Myra Adele Logan**—First African American woman elected to the American College of Surgeons; analyzed, refined X-ray techniques used to detect breast tumors in women.
- **Walter Massey**—Physicist and educator; professor of physics.
- **Elijah McCoy**—Mechanical engineer; patented more than 50 inventions.
- **Jessie Price**—Veterinarian microbiologist; researches the effects of bacterial infection on animals.
- **Madame C. J. Walker**—First self-made woman millionaire; invented a formula for straightening and grooming hair.

Charles Drew

Female Scientists
- **Irene Curie**—Nobel Prize winner; synthesized artificial radioactive elements.
- **Marie Curie**—Polish chemist and physicist in France, two-time Nobel Prize recipient; discovered radioactive elements such as polonium and radium in collaboration with her husband, Pierre Curie.
- **Amelia Earhart**—Pilot; pioneered long-distance flight.
- **Jane Goodall**—Zoologist; studied chimpanzees.
- **Elma Gonzalez**—Biologist; studies membranes and other small cell structures.
- **Mary Leakey**—Anthropologist, pieced together bone fragments to reveal human ancestry.
- **Barbara McClintock**—Geneticist, Nobel Prize winner; pioneered the idea of jumping genes.
- **Margaret Mead**—Anthropologist; studied primitive and other cultures.
- **Beatrix Potter**—Biologist, children's author; discovered that lichens are a symbiotic relationship of algae and fungi.
- **Rosaly Yalow**—Nobel Prize recipient; developed radio immunoassay, a technique for identifying and measuring very small amounts of trace materials.
- **Maria Elena Zavala**—Botanist; studies how naturally occurring chemicals protect plant cell from freezing.

Ernest Just

Other Minority Scientists
- **Alonzo Atencio**—Physician, researcher; studies how oxygen is transported to tissues.
- **Paul Chu**—Physicist; discovered a superconducting material in 1986.
- **John Hernandez**—Physicist at the University of North Carolina at Chapel Hill; specializes in chemical physics.
- **John B. Herrington**—Lieutenant Commander, U.S. Navy, Astronaut Candidate, NASA; Chickasaw Indian.
- **Roberto Merlin**—Physicist, native of Argentina; studies interaction of light with semiconductors.
- **Clifton Poodry**—Biologist; grew up on the Tonawanda Seneca Indian Reservation in western New York, studies developmental genetics in fruit flies at the University of California.
- **Eloy Rodrigues**—Chemist; studies chemicals in plants; expert on rubber plants.
- **Clifford Schumacher**—Theoretical physicist, Chippewa/Sioux Indian.
- **Chien-Shiung Wu**—designed and carried out an experiment on subatomic particles.

Madame C. J. Walker

HELPFUL HINTS

There are a number of things teachers can do to successfully implement multicultural education. One crucial step is to know the parents and the community resources, "especially people from various ethnic backgrounds who have science careers or hobbies" (Tolman and Hardy, 1995, 78). An excellent way to involve the community is to establish an international/multicultural day and invite students and their families to share their cultural heritage and ways of life. Teachers should emphasize that each family is unique and each family has its own cultural identity.

For students with language barriers, hands-on and minds-on activities in which students manipulate concrete materials can aid in the development of better process skills and the attainment of higher levels of science achievement. Using concrete experiences also helps develop new vocabulary. When writing on the board, overhead transparencies, or students' assignments, teachers should strive for legibility, especially for students whose native language is not English. Teachers can also help limited English proficiency students by having a dictionary in the student's native language available for reference. Teachers experiencing difficulty communicating with a student's parents should solicit the help of a community member fluent in the student's native language. During a test or quiz, teachers should allow more time for students not yet fluent in English.

A wide range of teaching strategies should be used. Teachers should learn about students' learning styles (hands-on, sequential, visual, auditory, group, individual, oral expressive, written expressive, or global) (Martin et al 1994). Visual and concrete demonstrations can be helpful when making a point or teaching a concept. The cooperative learning approach can foster cooperation among students from diverse backgrounds. Teachers can set an example of cooperation by complimenting students for their diverse views, praising them for their unique skills, and exercising understanding and compassion.

These suggestions are in keeping with the *National Science Education Standards,* which state that assessment practices must "be reviewed for the use of stereotypes" and "identify potential bias among subgroups" (NRC, 1996, 85). Furthermore, assessment tasks should "accommodate the needs of students with physical disabilities, learning disabilities, or limited English proficiency" and "be set in a variety of contexts, be engaging to students with different interests and experiences, and must not assume the perspective or experience of a particular gender, racial, or ethnic group" (NRC, 1996, 85).

EDUCATING EVERYONE

It is an established fact that our society is composed of distinct cultures. At present, many classrooms represent a mini United Nations where students of different backgrounds learn under the same roof. As teachers we should do all we can to help students experience success and to maximize their learning potential. We have the enormous responsibility to reach out to all students in our care. The well-being of a harmonious society depends on the successful coexistence and interdependence of its varied populations. As our society is increasingly and inevitably becoming diverse, this diversity, in essence, brings strength and poses certain challenges to our educational system. Last, it is important for educators to provide useful and productive opportunities for multicultural students to experience success in the classroom and in the larger society. ✧

S. Wali Abdi is an associate professor of science education, Department of Instruction, 404 Education Building, The University of Memphis, Memphis, TN 38152, e-mail: abdi.wali@coe.memphis.edu.

REFERENCES

"An NSTA Position Statement: Multicultural Science Education," (1991, October, November) *NSTA Reports!* 7.

Banks, J.A. Multicultural education: Characteristics and goals. In J.A. Banks and C.A. McGee Banks, eds. (1993.) *Multicultural Education: Issues and Perspectives.* Needham Heights, Mass.: Allyn and Bacon.

Martin, Jr., R.E., C. Sexton, K. Wagner, and J. Gerlovich. 1994. *Teaching Science for All Children.* Needham Heights, Mass.: Allyn and Bacon.

National Research Council. 1996. *National Science Education Standards.* Washington, D.C.: National Academy Press.

Tolman, M.N., and G.R. Hardy. 1995. Discovering Elementary Science: Method, Content and Problem-Solving Activities. Needham Heights, Mass.: Allyn and Bacon.

FOR FURTHER READING

Antonauris, G. 1989. Multicultural science. *The School Science Review* 70(252):97-100.

Bennett, D.I. 1995. *Comprehensive Multicultural Education: Theory and Practice* (3rd ed.). Needham Heights, Mass.: Allyn and Bacon.

Dunn, R., and K. Dunn. October 1975. Finding the best fit learning styles, teaching styles. *NASSP Bulletin* 59:37-49.

Hernandez, H. 1989. *Multicultural Education: A Teacher's Guide to Content and Process.* New York: Merrill/Macmillan.

MULTICULTURAL INTERNET SITE

www.tenet.edu/academia/multi.html Site includes links to African/African American, Asian/Asian American, Indigenous People, Latino, Native American, and general multicultural resources.

Teaching Essentials Economically

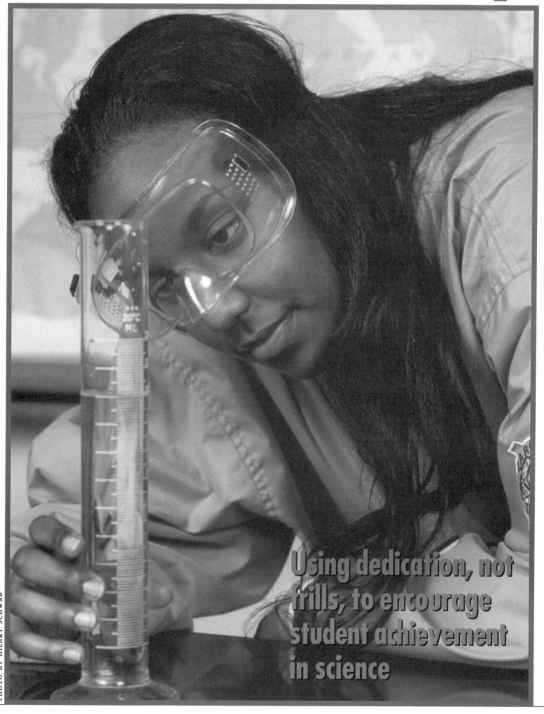

Using dedication, not frills, to encourage student achievement in science

THE VISION I HAVE FOR MY STUDENTS, who are mainly African American females, can be summarized in one word—success. I want them to experience success in an area of academia that once was off limits to them because of their race and sex. They need to know that access to science is not limited to the Albert Einsteins of the world but also is available to the Anitas, Mayas, and Angelas of the world. The thinking and reasoning skills they develop in my class may one day encourage them to pursue careers as doctors, teachers, lawyers, and scientists. Contagious enthusiasm, tenacity, and high expectations in science and higher education will replace once-held attitudes of frustration, pessimism, and hopelessness.

I envision my students returning home from college as successful professionals sharing new experiences and resources. This, I hope, will begin to break the cycle of poverty that hovers over our community.

OVERCOMING OBSTACLES

Attitudes and perceptions about science are powerful motivations that can either work for or against student achievement. Students who enjoy science and make high grades are more apt to do well and take advanced science courses. Similarly, students who dislike or fear science and doubt their own competencies are more likely to do poorly and stop taking science courses altogether by late high school (Kober, 1991).

To understand the negative attitudes many students in the southeast United States carry, it helps to understand the educational challenges here. Eudora Middle School is in a rural school system on the southern tip of Mississippi where the states of Arkansas, Mississippi, and Louisiana meet. The area, fondly called the Delta, is an agricultural area where cotton is still king and science seems to have little importance. Additionally, the Eudora public schools have an enrollment of 99 percent African Americans, the poorest and least educated demographic group in the tri-state area. The remaining 1 percent of students are of Hispanic and Caucasian descent.

Teaching science in Eudora is difficult because monetary resources are limited. For example, there is no formal lab in my classroom, and there is no running water in the classroom for science experiments. Students put their desks together to create lab tables because we cannot afford lab tables. One year, the school system was not able to furnish enough textbooks for classes. Obviously, it is hard to stay positive and motivated under these conditions. But it helps to remember my goal of nurturing a better appreciation of science so that one day my students can get science-related jobs and bring their skills back to the community.

I like teaching science in the Delta because, as an African American teacher who grew up here, it is easy for me to understand these particular students' fears and frustrations. I have experienced the challenges students face such as racial discrimination, economic hardship, and low-quality education. Mississippi, Louisiana, and Arkansas are among the worst-ranked states in the nation in education (Hovey et al, 1999).

When I began teaching, my biggest challenge was overcoming the negative feedback I received during my own educational experiences in the South. I knew if I had these struggles as a college-educated adult, less educated youngsters would be all the more challenged unless someone helped them before negative attitudes about science set in. When I was younger, my science teachers told me I would never understand science because I was not smart enough. While in graduate school, where professors delighted in my progress and encouraged me, I learned that I could, in fact, succeed in science. This experience taught me that anyone can succeed if given the right opportunities, support, and encouragement.

CHANGING ATTITUDES

I teach eighth grade students, ages 13 to 16. My students often have missed the early science opportunities offered in private schools and better-funded institutions. As a result of this disparity, they are behind in science before they enter middle school. Regardless, I try to teach them that they can learn and experience success in science. With dedication, they are realizing that they can take difficult science concepts and apply them to schoolwork as well as life situations.

Students are expected to develop skills in measurement, graphing, and problem solving that they can apply to scientific investigations they develop themselves. My goal for the unit is to have students demonstrate their understanding through participation in a school science fair. It takes 16 weeks of classroom instruction and individual work on science projects to complete the goal, and students are encouraged to do science fair projects at school to ensure participation.

I focus on the following objectives:

- Students must learn and review basic safety rules.
- Students must identify and review basic laboratory equipment and their functions.
- Students must construct graphs, tables, and charts from experiments.
- Students must apply the steps of experimentation when doing a science fair project.

To address the first objective, I introduce safety in the classroom. Some of the rules are: no horseplay, no eating in the classroom, goggles must be worn by students handling chemicals, hands must be washed after using chemicals, and equipment must be handled with care. Students are shown the nearest fire extinguisher, and a safety poster is pointed out with other safety rules.

The second objective is introduced as students are encouraged to examine instruments such as balances,

JOY R. DILLARD

scales, and pipettes that are laid out on a table. After examining the instruments, a handout is given that explains each one's purpose. The students are then required to draw each instrument and write down its function from the handout, and they practice using the instruments during the rest of the school year in different labs.

The third objective, for students to learn how to construct tables, charts, and graphs, is introduced through a demonstration of how to record experimental outcomes and measurements. Students practice making charts and tables from laboratory data they gather from measuring the growth of plants.

Students are then introduced to independent and dependent variables and graphing and are taught the basic types of graphs and the components of the graph. Finally, they are taught to plot points. To practice graphing, they are given data from experiments to graph, and they also have to graph the measurements of plant growth from the charts and tables they have made previously.

The final objective, for students to learn the steps of experimentation, is met through participation in a science fair project. Over several days, students are expected to design their experiments. I encourage them by telling them they are doing the same thing "real" scientists do every day.

Once students are ready to do their science fair projects, the school librarian and English teacher brief them on the mechanics of researching and assembling a term paper. They are given 25 ideas to consider and possibly research, but their ideas are generally more creative, such as "Will a Homemade Thermometer Measure Hot and Cold Temperatures?" and "Will Bleach Change the Shape of Containers?"

ASSESSMENT OF STUDENT LEARNING

Student progress is monitored with written tasks, homework, graphs and charts, journal writings, cooperative groups, peer assessment, and rubrics. In the first objective on safety and the second objective on the identification and use of laboratory equipment, the assessment is a laboratory safety contract that both parents and students read and sign. For more information on safety contracts, teachers may read "Contracting for Safety," in the September 1999 issue of *The Science Teacher* (Davidson, 1999). Students are given a safety test on which they have to distinguish good safety practices from bad ones, and everyone is required to score 80 percent or higher before participating in laboratory activities. Students are also assessed on their ability to identify the names and uses of instruments.

To meet objective two, each student must master basic measurement skills. For a final assessment on measurement, they make meter sticks out of construction paper. Then, they are assigned several items such as their textbook and the blackboard to measure.

For the assessment of objective three, students plant seeds and measure the heights of plants. The data is recorded in charts and tables and graphed over a three-week period. They also receive a daily grade on other graphs they have to read and interpret.

The fourth objective is assessed on how well students design their science fair experiments. Students are required to write lab reports that include all the steps of experimentation.

The science fair project is assessed by a rubric (Figure 1) that averages the scores for each category to determine the final project overall grade. The rubric is a good assessment tool because students have an opportunity to do well overall, even if they do not do well in a single area. Four categories are considered when assessing the project—student/teacher conferences, research papers, display boards, and journal writings.

The student/teacher conferences are assessed simply with a record of the total conferences held with each student. Students are required to initiate their own conferences. To assess this category, the total number of conferences is divided by the total number required for a particular grade.

To assess the research paper, certain criteria have to be followed that are based on the rubric. The paper is

FIGURE 1.
Modified scoring rubric.

100 points if check marks in all categories

Student/teacher conferences
—Four completed conferences

Research papers
—2.5 page report with information clearly stated, explaining independent and dependent variables and how they relate to the experiment
—Title page included with report
—Separate bibliography page included with heading
—Three sources arranged in alphabetical order on bibliography page

Display boards
—Display board with attractive title
—All six parts of experimental design written on typing paper
—All six parts of experimental design heading written in matching ink or typed
—All six parts of experimental design explained correctly and stated clearly
—Table, chart, or graph explained correctly
—Information on table, chart, and graph explained correctly
—Information on table, chart, and graph written neatly
—Photos glued neatly
—If photos are not used, a drawing is included
—Drawing is attractive, neat, and colored
—If drawing is on paper, it is glued down neatly

Journal writings
—Must average at least a 90 percent grade on journals

graded according to these criteria. If a student fails to include a particular criterion, as many as 10 points are taken from a total of 100 points until a final grade is obtained.

The assessment of the display board is also graded according to a rubric on which a total number of check marks are accumulated and then divided by the total number of checks required for a final grade on the display board. At the end of the project, all four categories are averaged for one final grade.

The results of using this type of assessment are tremendous. Last year, out of 86 students, only two did not participate. Twenty-two first-place ribbons were given to students averaging between 90 and 100 points. Thirty-two second-place ribbons were given to students averaging between 80 and 89 points. Eighteen third-place ribbons were given out to students averaging between 70 to 79 points. Only two people received honorable mentions for averaging below 70 points.

All journal writings used in the final project are peer assessed. Students are given a certain question to answer in their journals. For example, students have to define the word "control" and give an example of the control used in their project. Each journal is graded on a 100 percent scale, and the total number of journal grades is averaged for a final journal grade.

The results of the science fair show that the school's science program is improving. This was the first time in the history of the school that there was a 98 percent participation rate in a school-based event. My students demonstrated that anyone can do science if they are taught to organize data and logically approach and solve problems.

In the past, students at the Eudora Middle School had a poor record of science achievement. But this record is beginning to change as a result of the use of new methods that emphasize higher thinking skills and investigation instead of rote memorization. By experiencing success, my students have begun to reverse their negative attitudes—they believe they can achieve not only in science but also in life. ✧

Joy R. Dillard is a science teacher at Eudora Middle School, 210 Knox Street, Dermott, AR 71638.

REFERENCES

Davidson, A.B. 1999. Contracting for safety. *The Science Teacher* 66(6):36–39.

Hovey, K.A., H.A. Hovey, and H.A. Hovey. 1999. *CQ's State Fact Finder: Rankings Across America.* Washington, D.C.: Congressional Quarterly Books.

Kober, N. 1991. *Ed Talk: What We Know About Science Teaching and Learning.* Washington, D.C.: Council for Educational Development and Research.

Structured Observation

Charting student-teacher interactions to ensure equity in the classroom

ONE MAJOR ASPECT OF EFFECTIVE teaching is the attention paid to equity in the classroom. All students should receive equal attention from teachers, regardless of gender or race. Although the *National Science Education Standards* have called for the "provision of equitable opportunities for all students to learn science" (National Research Council, 1996), the reality is far from ideal.

As science teachers who have served as cooperating teachers for student teachers and worked collaboratively with university supervisors and researchers, we have developed a valuable tool for the assessment of teaching practice (Scantlebury et al, 1996). This tool quantifies the number and kinds of interactions between students and teachers. Structured observation of classroom events indicates whether or not a teacher is interacting equitably with each student in a classroom. The coding tool presented here is useful not only for training student teachers and mentoring novice teachers but also can be used to analyze the teaching practice of experienced teachers. The results from coding classroom interactions can effectively illustrate the actual classroom environment to teachers who, preoccupied with content, classroom management, or other issues, may not be aware of overall patterns in teaching practice.

The coding tool, originally developed for research purposes, was modified in order to be more appropriate for use in real classroom situations. It provides a pictorial representation of a teacher's interactions with students and provides numerical data from which calculations can be made and analyzed.

For more in-depth applications, the data may be used as a measure of change in interaction patterns, possibly by comparing observations over time. The data collected with the coding tool can provide simple and complex information by allowing the user to choose the appropriate level of specificity. For example, observers may separate the numerical data according to individual classroom populations. The actual manner in which the coding tool is used during an observation may vary—for instance, different symbols may be used, or the observer may shift to anecdotal notes if necessary. This tool is not meant to replace notes but rather to enhance the meaning of those notes.

Using the coding tool involves the quantitative collection of data on a seating chart. We have organized the guidelines for use of this coding tool in a progressive fashion, from the simplest coding of general interaction patterns to more complicated strategies that include any or all of the following groupings.

■ Interactions sorted according to types of academic questions.
■ Interactions sorted into academic and non-academic.
■ Interactions coded as teacher-initiated versus student-initiated.

BY ELLEN JOHNSON, BARBARA BORLESKE, SUSAN GLEASON, BAMBI BAILEY, AND KATHRYN SCANTLEBURY

In practice, it is probably easier to start with the more general observations and increase the complexity as one becomes more familiar with the coding tool.

BASIC CODING

Figure 1 shows a generalized seating chart that can be modified to reflect the actual layout of the classroom. Students should be identified on the chart by name, initials, or number. Other relevant information, such as sex, race, or ability level may also be included.

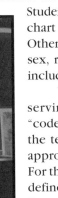

While in the classroom, the observing teacher should record or "code" all student interactions with the teacher by placing a mark in the appropriate box on the seating chart. For these purposes, an "interaction" is defined as an initiation and a response. For example, one interaction may involve the teacher asking, "What is a proton?" and the student answering, "I don't know." The observer may also decide to include interactions that include nonverbal responses, such as nodding. The coding procedure can be modified at the teacher's discretion, but it is important for the observer to be consistent within a given observation period. Global interactions addressed to the class as a whole or that involve "choral" responses on the part of the students may be noted in the margin or on a adjoining list of anecdotal comments. At the end of the coding period (which need not be an entire class period), teachers have a record of the pattern of the interactions that occurred during that time (Figure 2).

Certain aspects of the coding tool's use are meant to be flexible, such as who uses it, for how long, and how many observations are made. Obviously, the tool becomes easier to use when the observer has gained more experience with it, and the results are most meaningful when they reflect numerous hours of classroom observation.

CODING BY QUESTION TYPE

The coding tool can be modified to give specific information about the nature of each teacher-student interaction. For example, rather than using generic marks, the observing teacher can indicate whether an academic question is a knowledge-level (K) or upper-level (U) question. In general, knowledge-level questions are fact-based and involve recall, such as the use of vocabulary that has been previously discussed in the classroom. Upper-level questions are more conceptual in nature and may involve synthesis, problem solving, or critical-thinking skills. For example, if a definition was discussed or memorized from a vocabulary lesson, the question "What is a hermaphrodite?" would be knowledge level. On the other hand, if students were asked to name an example of a hermaphrodite without having been previously introduced to the concept, the question would be upper level. The distinction between these two types of academic interactions is best defined by the user of the coding tool, preferably with input from the teacher being observed.

ADDITIONAL INTERACTIONS

Obviously not all interactions that occur in a classroom are of an academic nature. The coding procedure can be adapted to reflect this. In addition to the different types of academic questions, the observer can also note procedural (P), disciplinary (D), and non-academic (N) interactions. Procedural questions have to do with the "nuts and bolts" of classroom agenda, such as "Do we have to write this in pen?"

Most disciplinary interactions are obvious enough to be designated as such, but occasionally the observer must judge whether an apparent academic question is also serving as a classroom management tool, for example, "Can you summarize what we've been talking about?" How to code such a question is up to the judgment of the observer, as long as the practice is consistent.

Observers should be aware that not all interactions are necessarily in a question-and-answer format. Students or teacher may initiate an interaction with a comment that may be on or off the topic. For example, "My father mixed Clorox and Comet and nearly had to go to the hospital because of the fumes," would be such an academic response if the discussion was on chemical reactions, toxic fumes, or chemistry in the home. Volunteered information that is not academic ("Did you know that it's my birthday?") can be coded as N. Figure 3 illustrates this type of coding. Double-coding (coding a single interaction more than once) should be avoided, however, as it will skew the number of interactions that involve a particular student.

As teachers strive to make their classrooms more "student-centered," the interactions that occur are initiated by teachers less and less frequently. It is desirable to account for this while observing classroom environments. One practice that works well is to divide each "seat" on the seating chart into two parts, labeling one Teacher to Student (T>S) and the other Student to Teacher (S>T) for teacher-initiated and student-initiated interactions, respectively.

When coding the class, if an interaction is initiated by the teacher, it is coded under T>S; if an interaction is initiated by a student, it is coded under S>T. Data should be separated according to T>S and S>T before making any calculations. Figure 4 illustrates this type of coding. The visual display of the coded seating chart alone can be dramatic, but is even more effective when the data is converted into percentages of total interactions as shown in Figure 5.

FIGURE 1.
Sample seating chart.

1.1	2.1	3.1	4.1	5.1	6.1
1.2	2.2	3.2	4.2	5.2	6.2
1.3	2.3	3.3	4.3	5.3	6.3
1.4	2.4	3.4	4.4	5.4	6.4
1.5	2.5	3.5	4.5	5.5	6.5
1.6	2.6	3.6	4.6	5.6	6.6

FIGURE 2.
Basic coding of student-teacher interactions.

1.1 CM ₥	2.1 MF //	3.1 MF	4.1 CF	5.1 MM /	6.1 CF //
1.2 CM	2.2 CF ₥ /	3.2 CM /	4.2	5.2 CF ////	6.2 CF //
1.3 CF ////	2.3 CM /	3.3 CM //	4.3 MM /	5.3 CF ///	6.3 CM ₥ ₥
1.4 CM /	2.4 CF ///	3.4 CM ////	4.4 MF	5.4 CF ///	6.4 CM //
1.5 CM ₥	2.5	3.5	4.5	5.5	6.5
1.6	2.6	3.6	4.6	5.6	6.6

REVIEWING THE DATA

Using either direct visual examination of the seating chart or calculations of percentages of total interactions, it is helpful to review the data as close to the observation period as possible, preferably in a conference with both the observer and the teacher being observed. Given the limited time allocated to professional development in many busy teaching schedules, this may not always be possible. In reviewing the data, teachers should note consistent patterns of interactions and question why these may occur. Some questions to consider include the following:

- Are the teacher's interaction patterns consistent over time?
- Are the interaction patterns equitable or appropriate? If not, what accounts for the differences?
- Is the classroom environment equitable for both genders? For students of all ethnicities? For those of different ability levels? For individual students?
- What nonacademic issues seem to influence who interacts with the teacher? What are the patterns of procedural or disciplinary interactions?
- Is every student's understanding being assessed during the lesson? To the same depth?

FIGURE 3.
Coding indicating type of interaction: K=knowledge-level, U=upper-level, P=procedural, N=nonacademic, D=disciplinary

1.1 CM KPKNK	2.1 MF KK	3.1 MF	4.1 CF	5.1 MM P	6.1 CF KD
1.2 CM	2.2 CF PPPKPK	3.2 CM K	4.2	5.2 CF PDDK	6.2 CF DD
1.3 CF KKPP	2.3 CM K	3.3 CM KK	4.3 MM U	5.3 CF KPP	6.3 CM KKUNN KKKUK
1.4 CM K	2.4 CF KKU	3.4 CM KKUK	4.4 MF	5.4 CF DDN	6.4 CM DN
1.5 CM KKPPP	2.5	3.5	4.5	5.5	6.5
1.6	2.6	3.6	4.6	5.6	6.6

FIGURE 4.
Coding indicating whether interaction is teacher-initiated or student-initiated.

1.1 CM KPK / NK	2.1 MF / KK	3.1 MF	4.1 CF	5.1 MM / P	6.1 CF KD /
1.2 CM /	2.2 CF PPPK / PK	3.2 CM / K	4.2	5.2 CF PDD / K	6.2 CF / DD
1.3 CF KK / PP	2.3 CM / K	3.3 CM / KK	4.3 MM / U	5.3 CF K / PP	6.3 CM KKUK UU / K NN
1.4 CM / K	2.4 CF / KK U	3.4 CM / KKU K	4.4 MF	5.4 CF DD / N	6.4 CM / D N
1.5 CM KKP / PP	2.5	3.5	4.5	5.5	6.5
1.6	2.6	3.6	4.6	5.6	6.6 T>S / S>T

MAKING CHANGES

During our three-year use of the coding technique, we found that preservice student teachers and inservice teachers alike employed a number of practical and sometimes creative changes in teaching practice in response to their interaction data. These included some methods already recognized as good teaching practice, such as moving about the room while addressing students to create differing focal points in the classroom. Reviewing the interaction data was often an effective way to get teachers to pay more attention to these techniques.

Some teachers came up with solutions such as keeping checklists of which students were involved in interactions over a given period of time. Others shuffled note cards with students' names in order to assure random interactions. In some cases, student teachers trained cooperating teachers in the use of the tool in order to get more feedback for the student teachers. In less traditional classrooms where the use of cooperative groups was well established, coding data gave an objective measure that was useful for reassigning students to groups based on ability level or learning style. We also noticed that teachers incorporated new levels of interactions (student to student) within their classrooms, such as allowing a student to "toss the ball" to another student in order to have that student expand on or explain a concept.

In general, we found that science teachers were receptive to the relatively quantitative nature of the process. In many cases, the tool provided a nonthreatening, objective basis for communication between cooperating teacher and student teacher and helped to avoid situations where criticism was stressful and uncomfortable. The observation tool also responds to the call for alternative modes of teacher evaluation (Gitlin and Smyth, 1990) in order to increase equity in the classroom. ✧

Ellen Johnson is a visiting professor at the University of Delaware on leave from Concord High School, 2501 Ebright Road, Wilmington, DE 19810; e-mail: emjohnso@udel.edu. Barbara Borleske is a chemistry teacher at John Dickinson High School, 1801 Milltown Road, Wilmington, DE 19808; e-mail: bborlesk@udel.edu. Susan Gleason is a chemistry teacher at Middletown High School, 508 S. Broad Street, Middletown, DE 19709; e-mail: sglea@udel.edu. Bambi Bailey is an assistant professor, Department of Curriculum and Instruction, Texas A & M International University, Laredo, TX 78041; e-mail: bbailey@tamiu.edu. Kathryn Scantlebury is an associate professor at the University of Delaware, Department of Chemistry and Biochemistry, Newark, DE 19716; e-mail: kscantle@udel.edu.

NOTE

This material is based on work supported by a grant from the National Science Foundation (Grant No. HRD 9450022). Any findings, opinions, conclusions, and/or recommendations expressed in this article are those of the authors and do not necessarily reflect those of the National Science Foundation.

REFERENCES

Gitlin, A., and J. Smyth. 1990. Toward educative forms of teacher evaluation. *Educational Theory* 40(1):83-94.

National Research Council. 1996. *National Science Education Standards*. Washington, D.C.: National Academy of Sciences.

Scantlebury, K., E. Johnson, S. Lykens, R. Clements, S. Gleason, and R. Lewis. 1996. Beginning the cycle of equitable teaching: The pivotal role of cooperating teachers. *Research in Science Education* 26(3):271-282

FIGURE 5.

Percentage of total interactions during sampling period broken down by gender and race.

	MM% of Total Interactions	CM% of Total Interactions	MF% of Total Interactions	CF% of Total Interactions
T>S P	0.0	6.7	0.0	8.3
T>S K	0.0	28.3	0.0	13.3
T>S U	0.0	6.7	0.0	0.0
T>S D	0.0	1.7	0.0	8.3
T>S N	0.0	1.7	0.0	0.0
S>T P	1.7	0.0	0.0	6.7
S>T K	0.0	1.7	3.3	1.7
S>T U	1.7	0.0	0.0	1.7
S>T D	0.0	0.0	0.0	0.0
S>T N	0.0	5.0	0.0	1.7
Total	3.4	51.8	3.3	41.7
Key	MM=Minority Male	CM=Caucasian Male	MF=Minority Female	CF=Causasian Female

Notable Women

Teaching students to value women's contributions to science

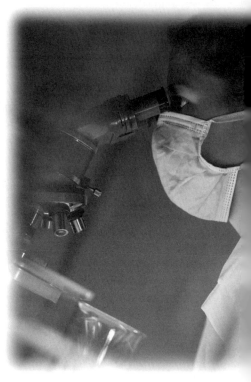

ASK MIDDLE OR HIGH SCHOOL STUDENTS to name a scientist, and they will most likely name Albert Einstein, Louis Pasteur, or Thomas Edison. They might name Marie Curie, but probably not. Students instructed to draw a scientist seldom draw a woman. The fact is, most people do not picture women when they think of scientists. Not only does this sad fact cloud our knowledge of women's contributions to science throughout history but it also subconsciously maintains the idea that science is men's work.

If we want female students to have a chance to grow into successful and renowned scientists, we must make a concerted effort to provide them with role models. If we want male students to grow into men with great respect for women, both personally and professionally, we should show them that adults value women's contributions. A science classroom is the perfect forum for making a positive contribution to both goals.

MODEL SCIENTISTS

Research over the last 15 to 20 years has shown that as female students reach junior high and high school, their willingness to take rigorous science courses steadily decreases. Although the situation is improving, young women are continuing to enter classrooms in which they face bias against enrolling in higher math and science courses. Whether this bias is overt discrimination or a covert lack of support for their progress, these female students feel insufficient encouragement at the time it is needed most.

A cursory look at many textbooks, particularly older editions, shows considerable inequality in the number of photographs of women engaged in science. Examinations of the posters in many classrooms reveal the same disparity. An almost consistent use of the pronoun "he" when referring to scientists also communicates that science is men's work. These messages speak volumes to all students. Only diligent attention to combating this problem will bring about change.

After mulling over this situation for some time, I developed a project focusing on women in science for my high school biology students. With the constraints of few resources and little time, the project had to be quick but meaningful. I also wanted to combine science with another discipline and decided on business. The final outcome was a project in which students research a woman scientist, create a résumé for her, and deliver an oral presentation about her life.

Each group within a class period must choose a different scientist. Students are not permitted to check out books from the school library because a large number of students usually work on the same project. Résumés are graded on a comparison basis against students researching the same scientist in other classes and in previous years. Students may use any factual sources available to them.

Four days are normally adequate for the project, including research and oral presentations. On day one I introduce the project, distribute handouts, assign the students to groups of four, and send them to the library to get started on their research. I provide a list of women scientists from which students may choose. (Figure 1 is a list of women from which to select. I recommend putting

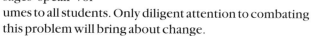
CINDY L. F. ZACKS

an asterisk next to the women who may be easier to research and a check mark next to those who may be found on the Internet.) I also provide project instructions (Figure 2), Internet sites, and an example of a résumé I created for one of my role models, Rachel Carson (page 52). I explain to students that the résumés must follow the format of my example and cover six topics: background information, education, work experience, professional skills, awards and achievements, and activities and interests. On a separate piece of paper, students are required to write a few paragraphs citing examples of prejudice faced by the group's scientist and discuss discrimination against women in general.

Students in each group must divide the work equally, each taking responsibility for two sections. They write their name in the left-hand margin in front of the sections for which they took responsibility and write the sections they researched on the final draft. While researching the scientist,

FIGURE 1.
Women scientists.

- Annie Jump Cannon–astronomer
- Rachel Carson–ecologist
- Williamina Paton Stevens Flemming–astronomer
- Catherine Furbish–botanist
- Libbie Henrietta Hyman–zoologist
- Clara Barton–founder of the Red Cross
- Ruth Fulton Benedict–social anthropologist
- Elizabeth Blackwell–first American female doctor
- Maria Goeppert Mayer–physicist
- Henrietta Swan Leavitt–astronomer
- Mary Watson Whitney–astronomer
- Eileen Collins–astronaut
- Marie Curie–chemist
- Mae Jemison–astronaut/physician
- Rosalind Franklin–crystallographer
- Jane Goodall–anthropologist/primatologist
- Shannon W. Lucid–astronaut
- Florence Augusta Merriam Bailey–ornithologist
- Dorothy Hodgkin–crystallographer
- Elizabeth Gertrude Knight Britton–botanist
- Cornelia Maria Clapp–zoologist
- Hypatia–mathematician/astronomer
- Anna Botsford Comstock–naturalist
- Katherine Esau–botanist
- Ruth Sager–geneticist
- Stephanie Kwolek–chemist
- Rosalyn Sussman Yalow–medical physicist
- Ellen Russell Emerson–ethnologist
- Mary Leakey–anthropologist/archeologist
- Margaret E. Knight–inventor
- Zelia Maria Magdalena Nuttall–archeologist
- Mary Breckinridge–nurse/midwife
- Sucy Beaman Hobbs Taylor–dentist
- Florence Rena Sabin–anatomist
- Ynes Mexia–botanist
- Ann Haven Morgan–water ecologist
- Dorothy McClendon–biochemist
- Dale Emeagwali–biologist
- Marie Maynard Daly–chemist
- Katherine G. Johnson–physicist
- June Bacon-Bercey–meteorologist
- Lee Anne Martinez–ecologist
- Sonia Ortega–marine biologist

- Maria Elena Zavala–botanist
- Osa Helen Leighty Johnson–explorer
- Margaret Mead–anthropologist
- Lise Meitner–physicist
- Dian Fossey–anthropologist/primatologist
- Sacagawea–naturalist
- Maria Mitchell–astronomy
- Mary Morris Vaux Walcott–naturalist
- Florence Nightingale–nurse
- Hattie Elizabeth Alexander–microbiologist
- Hazel Gladys Bishop–chemist
- Gerty Theresa Radnitz Cori–biochemist
- Gertrude Belle Elion–biochemist/pharmacologist
- Alice Hamilton–pathologist
- Grace Brewster Murray Hopper–computer science
- Ana Maria Azzarolo–physiologist
- Barbara McClintock–geneticist
- Chien-shiung Wu–nuclear physicist
- Rita Levi-Montalcini–neurologist
- Margaret Warner Morley–biologist
- Nettie Maria Stevens–biologist
- Katherine Siva Saubel–ethnologist
- Lucy Myers Wright Mitchell–archeologist
- Alice Cunningham Fletcher–anthropologist/ethnologist
- Virginia Apgar–physician
- Helen Broinowski Caldicott–physician/anti-nuclear activist
- Elizabeth Stern–pathologist
- Mathilde Krim–medical researcher
- Edith Patch–entomologist
- Alice Eastwood–botanist
- Ruth Ella Moore–biochemist
- Roger Arliner Young–zoologist
- Shirley Ann Jackson–physicist
- Marguerite Thomas Williams–geologist
- Leticia Marquez-Magana–molecular biologist
- Elvia Niebla–soil scientist
- Lydia Villa-Komaroff–biologist
- Martha Zuniga–biologist

Internet sites:
www.greatwomen.org/index
www.nobel.se
www.etdc.wednet.edu/equity
www.sacnas.org

each student is required to take a page of notes that, along with the written portions on the final résumé, prove participation in the project.

At the conclusion of three class periods spent in the library doing research, final résumés (one for each group) and individual notes are due. Three days in the library are sufficient for completion of this project. As long as students have completed a résumé and followed directions, they earn full credit for the written assignment.

Within the following week, I dedicate one class period to oral presentations. Each person must present research, which results in all group members speaking an equal amount of time. The résumés can be used as guides, but students are not allowed to read the information verbatim. As long as students complete their presentations in a professional manner, 50 points are earned for the oral reports. To ensure audience participation, all students are expected to take notes during each presentation. These are collected, and points are earned based on effort. Students who are absent on any day of the in-class research time are required to complete the project alone, including the oral presentation. If they are absent for the oral report, they are required to complete a research paper in addition to a résumé.

FUTURE BENEFITS

While this sounds like a whirlwind introduction to the topic, it often is the first time most of my students have studied women scientists. During the course of research, students often make comments regarding discrimination women have faced throughout history and condemn such treatment. Female students often thank me for the opportunity to explore career options they did not know existed.

Prior to a change in scheduling, this project was combined with a "Careers in Science" unit. When my school changed to the trimester system, I was forced to delete most units not absolutely essential to biology education. I was unwilling to cut my "Women in Science" project, however, because it is essential to students' futures. If I were able to expand this unit again, students could construct their own "professional" résumés after completing their library project. I would also invite local women scientists to speak with my classes. Their inclusion would give students tangible examples of women working in scientific fields, and these presenters could be the seeds for a mentorship program.

Before embarking on this project, teachers should review their school's library for appropriate resources. Most likely such resources will be scant, and supplemental research materials will have to be purchased. The following books are particularly useful: *Extraordinary Women Scientists* (Stille, 1995), *Women in the Field, Pioneer Women Naturalists* (Myers Bonta, 1991), *Women in Science* (Bailey Ogilvie, 1986), *Women of Science, Righting the Record* (Kass-Simon, 1990), and *Nobel Prize Women in Science* (Bertsch McGrayne, 1993). I have found the National Women's History Project (7738 Bell Road, Windsor, CA 95492; *www.nwhp.org*) to be the best source of materials. ✧

Cindy L.F. Zacks is a biology/field ecology teacher at Yucca Valley High School, 7600 Sage Avenue, Yucca Valley, CA 92284; e-mail: zacks.d.and.c@thegrid.net.

REFERENCES

Bailey Ogilvie, M. 1986. *Women in Science.* Cambridge, Mass.: The MIT Press.

Bertsch McGrayne, S. 1993. *Nobel Prize Women in Science.* New York: Birch Lane Press Book.

Kass-Simon, G., ed. 1990. *Women in Science, Righting the Record.* Bloomington, Ind.: Indiana University Press.

Myers Bonta, M. 1991. *Women in the Field, Pioneering Women Naturalists.* College Station, Tex.: Texas A&M Press.

Stille, D. R. 1995. *Extraordinary Women Scientists.* Chicago, Ill.: Children's Press Inc.

FIGURE 2.

Guidelines for student reports

Personal Information—You must include her dates and locations of birth and death. You must also include her profession(s).

Education—You must include bachelor's degree, master's degree, and Ph.D. Each entry must include date of completion, school name, and subject of degree(s). If your scientist had no formal education then you must explain in detail how she was educated and how she became so well known in her field of study.

Work Experience—Must include dates of employment, location, and job descriptions. You must have at least four entries. For some women, unpaid jobs are acceptable.

Professional Skills—List and give examples of at least four skills your scientist has/had that helped her to be successful.

Awards and Achievements—Must include dates, name of award, description of activity leading to award, description of achievement, and why you consider it an achievement. You must have at least four entries.

Activities and Interests—What does/did your scientist like doing? You must have at least four entries. On separate page, answer the following questions:

1. As a woman, what obstacles did your scientist have to overcome to succeed in the scientific world? Be specific.
2. What are some examples of prejudice against women, in general, that you found while researching your scientist? Give at least two examples. You may have to discuss this with other groups to get enough ideas.

Rachel Louise Carson

Photo: Rachel Carson on Hawk Mountain, Pennsylvania, 1945 watching migrating hawks. Photo by Shirley A. Briggs. Used by permission of Rachel Carson History Project.

Date of Birth: May 27, 1907
Date of Death: April 14, 1964

Professions: Rachel Carson was a writer, marine biologist, and ecologist.

Education:
Bachelor of Arts: Science—1929
Pennsylvania College for Women
Master of Arts: Zoology—1932
Johns Hopkins University

Work Experience:

1937 to 1964 *Science Writer:* Rachel Carson wrote and published many books about natural history. In 1962 she published her most famous book *Silent Spring* amid much controversy. She also published *The Edge of the Sea* in 1955, *The Sea Around Us* in 1951, and *Under the Sea* in 1941.

1940 to 1951 *Chief Publications Editor:* (U.S. Fish and Wildlife Service) Wrote, edited, and reviewed all publications prepared by Fish and Wildlife Service personnel.

1935 to 1940 *Junior Aquatic Biologist:* (U.S. Bureau of Fisheries) Studied and researched aquatic organisms. One of the first two women ever hired for a scientific position with the bureau.

1932 to 1935 *Zoologist:* (Johns Hopkins University) Studied and researched living organisms.

Professional Skills:
Writing: books, manuals, and magazine articles
Public speaking: teaching, news broadcasts, and television interviews
Research: scientific and political issues
Editing: reviewing articles and documents for publication

Awards and Achievements:

1980 Awarded the Presidential Medal of Freedom for work as a writer (after death)
1962 Elected to the American Academy of Arts and Letters for her book *Silent Spring*
1951 Elected to the British Royal Society for Literature for *The Sea Around Us*
1951 Awarded the National Book Award for her book *The Sea Around Us*

Activities and Interests:
Gardening, reading, writing, marine biology

Language Diversity & Science

Science for limited English proficiency students

EVERY DAY, MORE THAN 6 MILLION students come to school from homes in which a native language other than English is spoken. Nearly every teacher in every state has at least some contact with students whose native language is not English. Many of these students have emigrated from Asia, Africa, the Middle East, and Europe and thus are dealing not only with the demands of American education, but also with the challenges of a new culture and a new language.

The fact that limited English proficiency students face special problems is important for all teachers, but especially for science teachers. In the workplace of today and tomorrow, a more-than-nodding acquaintance with the technical world will be required, and science is the freeway to the technological disciplines, both in college and beyond. We cannot consign nonnative speakers to the sidelines, where they will be underemployed or unemployable. But how do we impart the complexities of science to children who already are struggling just to comprehend the nuances of their second language?

Research literature is virtually silent on the most effective methods for teaching science to such students. Reviewing science methods textbooks published since 1980, we discovered that in the nearly 8,000 pages written by and for science educators, fewer than 100 pages focus on teaching science to language minority students. Faced with such a dearth of information, many teachers have become discouraged and disheartened. For lack of any alternative, they often have fallen back on the "sink or swim" approach.

From 1990 to 1995, however, we instituted a variety of interdisciplinary research studies under the auspices of The National Center for Science Teaching and Learning. We surveyed pre-service teachers in elementary and secondary science methodology classes in Florida, California, Rhode Island, and Ohio about their concerns related to teaching students with limited English proficiency. In addition, we made long-term observations in two settings, studying 42 fourth- and fifth-grade students in a Spanish language immersion classroom at a Columbus, Ohio, magnet elementary school and studying 12 to 16 tenth-graders in a Santa Cruz, California, high school that houses Spanish-speaking students at risk of failure.

From these studies, which involved gathering and interpreting data from videotapes, interviews, surveys, classroom-based research, and second language acquisition databases, we identified several classroom concerns

BY ELIZABETH BERNHARDT, GRETCHEN HIRSCH, ANNELA TEEMANT, AND MARISOL RODRÍGUEZ-MUÑOZ

and some strategies for meeting the needs of limited English proficiency students. The following questions highlight a few of the major concerns.

Why do school systems send students into the science classroom before they are at least moderately fluent in English?

Delaying science instruction until children are fluent in spoken English dooms them to academic failure. If, as is commonly agreed, it takes between six and eight years for nonnative students to become fluent, today's first grader would receive no science instruction until age 12 to 14. From a science perspective, that is years too late. Research indicates that if students have not had a positive science experience by the fourth grade, they are probably lost to the sciences for good. If we wait to send limited English proficiency students to science class, they will never be ready for, or get, the science they need.

Additionally, in the science classrooms we studied, we discovered students who could understand far more than they could articulate. Take the concept of condensation, for example. In our research, we encountered students who drew painstaking diagrams of the concept and labeled them correctly in their native languages. It was clear students already had complete understanding of the process. They simply needed to learn to read, pronounce, and spell the term in English.

Paradoxically, simple terms such as *force* may prove more difficult for second language learners because the word has both a common and a scientific meaning. A monolingual student probably knows the common meaning; the multilingual student might not, and he or she will have to learn and understand both meanings.

How do I know that my students understand a concept when their English often is so garbled?

It is important to realize that a wrong word or a wrong tense does not necessarily mean a wrong concept. For example, students pass through predictable stages in the use of verbs (Larsen-Freeman and Long, 1991). They commonly learn irregular verbs, such as *think-thought* or *know-knew*, first. They then move on to the regular verbs, such as *walk-walked*. Because past tense and other, more complex tenses may be very confusing for them, they begin to overcompensate. When they attempt to formulate a prediction or state an observation, they may say or write such things as, "I had previously saw" or "I did counted." These types of errors are, in fact, good indicators that the students are learning a significant amount of English and should be applauded as signs of accomplishment.

The bilingual classroom is different from the monolingual setting. Bilingual students need more avenues of instruction, and teachers must devise more ways to check for understanding. These strategies are called on-line adjustments, and the most effective adjustments occur when teachers interact intensively with their second language students during labs or lectures.

Ask questions to see if the students understood what you said or to clarify your understanding of what they said. Yes or no questions such as, "Did you measure the amount of water in milliliters?" rather than questions such as, "What unit of measurement did you use?" give them a choice between two alternatives and makes it easier for them to respond.

Break down difficult ideas into more understandable segments. Pause now and then to give students a chance to catch up with and process your words. Stress the main word or idea of a sentence. Use synonyms for important words or simply repeat an idea in different words. Do not rely solely on the spoken word. Support your presentation by writing key words on the blackboard or providing a handout students can use to follow the discussion.

Provide a variety of opportunities for your second language students to read and write. If they have chances to read-and-do or write-and-do, they often can demonstrate understanding without the risk of speech—a risk that sometimes can seem overwhelming. Of course, these on-line student-teacher negotiations may be equally beneficial for monolingual students who are having trouble grasping science content.

Would it not be easier just to simplify the materials until these students got caught up with the rest of the class?

Second language students do not need "watered down" materials. Simplification does not always enhance comprehension. Say, for example, your class is studying the solar system. If you oversimplify the grammar and

> *If students have not had a positive science experience by the fourth grade, they are probably lost to the sciences for good.*

> *Second language students do not need "watered down" materials. Simplification does not always enhance comprehension.*

vocabulary, you deprive the students of the very words and linguistic structure they need to understand the science. Nonetheless, you may need to extend your explanations for second language students.

It is critical not to confuse modification (making on-line adjustments) with simplification. In fact, second language learners do much better with elaborated rather than simplified language. Giving them more than one word or example gives them more than one path to understanding a concept. For example, we saw second language students struggling with the distinction between *rotation* and *revolution* of the planets. One effective way to get the concept across is to have students role-play the planets as you or another student describes what is happening. Following the role play with a computer animation activity or a handout helps cement the concept. Making science comprehensible to second language students involves modifying the science content to provide a rich mix of reading, writing, listening, and speaking experiences.

I have some students who do not say anything at all. What do I do about them?

Learning a language is a psychological as well as a linguistic event. Language learning often begins in silence, as students listen and absorb as much as they can from the environment. They begin to speak when they feel more comfortable. For some students, fear of correction—loss of face—is a powerful incentive to keep quiet. For others, learning the new language may feel like a betrayal of their home culture and language. Once these students begin to open up, it is better to focus initially on *what* they say rather than *how* they say it. Modeling the proper word rather than correcting the student publicly may be useful when he or she is just beginning to speak.

All students need the opportunity to demonstrate their knowledge using the language skill—speaking or writing—with which they are most comfortable. A student who is nervous about her speaking ability may do very poorly in a lab demonstration. If you allow her to write, however, you may discover she is quite knowledgeable about the content. Her written product reflects her actual comprehension rather than her test anxiety. Reading, standardized tests, and performance assessment should all be taken into account when grading second language students.

Where can I go to learn more about other strategies for helping my language minority students?

ESOL professionals can help you establish language goals, while you, as the science expert, take charge of the content-area objectives. Science and language teachers need to work together to assure the academic success of the second language students they teach. ✧

Elizabeth Bernhardt is director of the Stanford Language Center, Stanford University, Building 320, Stanford, CA 94305; Gretchen Hirsch is associate editor at the National Center for Science Teaching and Learning, 1929 Kenny Road, Columbus, OH 43210; Annela Teemant is a graduate research associate at The Ohio State University, 249 Arps, 1945 N. High Street, Columbus, OH 43210; and Marisol Rodríguez-Muñoz is a science teacher at Gladstone Elementary School, 1965 Gladstone Ave., Columbus, OH 43203.

REFERENCES

Bernhardt, E.B. 1994. A content analysis of reading methods texts: What are we told about the nonnative speakers of English? *Journal of Reading Behavior* 26(2):159–189.

Bernhardt, E.B. 1995. Science education and the bilingual learner. Unpublished manuscript.

Larsen-Freeman, D. and M.H. Long. 1991. *An Introduction to Second Language Acquisition Research.* New York: Longman.

Meaningful

TRADITIONALLY, WE ASSUME STUDENTS will understand and learn a curriculum that is presented in a scientifically appropriate way. However, this assumption is valid less often than it used to be (Lee and Fradd, 1998) because the percentage of monolingual English-speaking students in the United States has decreased during the past decade while the number of students learning English as a second language has increased. This demographic shift requires teachers to address English language deficiencies when teaching science.

Integrating English language instruction with science represents a new role for many of us who are accustomed to teaching only science. To accommodate this additional responsibility, we need to examine effective instructional practices for making science instruction understandable and meaningful for English Language Learners (ELLs).

SHELTERED INSTRUCTION

Sheltered instruction is an important innovation in the education of ELLs (Faltis, 1993). In a sheltered science class, teachers use specific strategies to teach science in ways all students can understand while at the same time employ techniques that promote English language development. Some teachers see sheltered instruction as nothing more (or less) than good teaching because it incorporates many of the strategies found in high quality non-sheltered instruction. Sheltered classes, however, are distinguished by careful attention to students' needs related to learning another language.

The sheltered instruction model integrates science objectives and language development objectives, providing instruction that meets the unique needs of ELLs enrolled in grade-level content courses (Short and Echevarria, 1998). Taught effectively, grade-level sheltered instruction teaches students the same content they would learn in any science course and encourages them to use the same abilities and skills that are required in other courses. This type of instruction is not meant to be a watered-down version of other science courses. English use is constantly modulated or negotiated, and content information is made comprehensible through the use of visual aides such as pictures, models, demonstrations, and graphic organizers (Echevarria and Graves, 1998).

ALAN COLBURN AND JANA ECHEVARRIA

Lessons

All students benefit from integrating English with science

Sheltered instruction is both similar to and different from hands-on, open-ended teaching. The following example of a science activity, which introduces students to the concepts of density and buoyancy, illustrates the comparison.

TEACHING BY DEMONSTRATION

As students file into class, they see a half-filled aquarium at the front of the room. The teacher starts class by telling students to watch carefully as she places an orange into the tank of water. The orange floats. "Now watch," she announces, placing another orange (this one peeled) beside the first orange. "What do you notice?" she asks the class. A student responds by saying one orange is floating and the other sank to the bottom.

"Why do you think that is?" she continues. "Take a moment to think about it, then discuss it with the person sitting next to you." After pausing 30 or 40 seconds she calls out "So, what do you think?" Two or three students offer tentative explanations, including one about the orange peel acting as a life vest for the rest of the orange. "The purpose of this unit is for you to better understand why some things float while others sink. Before we are done, you will be able to calculate and predict whether something is buoyant enough to float," the teacher explains.

In the next part of the activity, students use florist's clay or something similarly water insoluble to make a small "boat" and see if it floats. Working in pairs, students record in their notebooks small drawings of the clay shapes they create, the extent to which the clay boats float, and their thoughts after each trial.

Switching from clay to aluminum foil, which is easier and cleaner to use, the teacher adds a quantitative element to the lesson. Students must make an aluminum foil boat, calculate its volume, and determine the maximum mass the boat will hold before sinking. The teacher asks students to record the mass of the loaded boat and find the mass-to-volume ratio. Students repeat their work with different-sized boats, and then the class pools its data, drawing a table on the overhead projector with masses and volumes in each column. The data will be used to make a graph with maximum mass held by a boat on the y-axis and the boat's volume on the x-axis.

Students who finish early calculate the masses for different volumes of water. This information also is entered onto the graph in a different color. This information lets students visually compare similarities between the mass-to-volume ratio for objects that float and the same ratio for water. This part of the activity also helps students see that the mass-to-volume ratio (density) of a substance (water) does not change even though the mass or volume itself may change. The concept is difficult for many students to grasp.

As students work, the teacher may comment, "Tell me about what you are doing," and "What are you thinking at this point?" The teacher listens carefully, paraphrases students, asks for clarification, and follows up with questions or statements based on what students say. She tries to get a better understanding of students' beliefs about why things float and sink and to help the students themselves better understand their belief.

The teacher continues asking students questions in which they clarify what they are doing ("So, tell me if this is right: you are saying that the more surface touching the water, the more mass your boat can hold before sinking. Is that right?"), test their ideas ("What could you do to be more certain about what you are saying?"), or make a prediction ("What do you think would happen if you made a boat with deep sides but little surface on the bottom?").

After most students have generated data about the maximum loads their boat can hold (making sure to also include the mass of the boat itself), the activity continues. With the class's graph available on an overhead transparency (and each student having made a copy of the graph), it is time for a brief lecture explaining the relationship between sinking, floating, and the mass-to-volume ratio (density). The teacher also assigns the class to read the section in the textbook discussing density.

On the next day, the students and teacher begin working through problems to which students apply their knowledge, make predictions about how much mass and volume of materials various real boats can hold, and begin investigations to quantitatively compare buoyancy in salt water versus freshwater.

ANALYZING THE LESSON FROM A SHELTERED PERSPECTIVE

Some aspects of this lesson would be beneficial for ELLs, while other aspects would be quite problematic for them. In brief, learning is enhanced in this lesson when the teacher provides visual representations of verbal information and when students have the opportunity to explore academic concepts through the use of hands-on materials. Areas of difficulty include the teacher using terms that may be unfamiliar to students, such terms as calculate, predict, buoyant, ratio, mass, and density.

Sheltered lessons are based on activities that are made meaningful when they provide students with hands-on materials with which to practice using new content knowledge. In this lesson, the concepts presented would be quite difficult to grasp if presented only in a lecture format. The teacher made the concepts meaningful by using hands-on materials, allowing students to "see" concepts such as buoyancy, mass, and volume.

Meaningful activities provide students with hands-on materials with which to practice using new content knowledge.

Sheltered activities integrate lesson knowledge and concepts with opportunities to practice using English by reading, writing, listening, and speaking. It is critical for ELLs to have sufficient opportunities to use academic English in a variety of ways. Instead of simply demonstrating lessons to the class, the teacher gave students the opportunity to work together not only listening but also talking, writing, and reading about academic concepts and language associated with the lesson. Working in pairs and small groups facilitates such opportunities.

Interaction is encouraged during sheltered lessons. Sheltered classes provide frequent opportunities for interaction and discussion between the teacher and students, as well as among students. The teacher consistently provides sufficient wait time for student responses and encourages elaborated comments about lesson concepts. ELLs require more time to process new vocabulary, so teachers should be aware of the need for wait time and the benefit of patient encouragement when students participate in discussions.

Most frequently, fluent English speakers will answer questions posed to the class. To be sure all students participate, especially those with limited English proficiency, the teacher should also use group response techniques. For example, she could say, "If you think the orange peel served as a life vest, hold up one finger. If you think there is another explanation, hold up two fingers." With a quick glance around the room, the teacher can see how many students are on target and determine if more explanation or clarification is necessary.

Sheltered lessons make extensive use of supplemental materials. Lessons are made clear and meaningful to ELLs through the use of such materials as graphs, models, illustrations, demonstrations, and outlined presentations on the overhead projector. ELLs benefit tremendously from the supplemental materials used in activities like this lesson. However, the teacher could do other things to make this lesson more effective for students learning English.

During the early part of the lesson, students are told what to record in their notebooks. In addition to telling, the teacher should also illustrate the point so students know exactly what they are expected to do. For instance, while doing the boat activity, the teacher could draw three columns on an overhead projector transparency. Then, the teacher could shape clay into a form, drop it into the water, and draw the shape (elongated) in the first column. In the second column the teacher could write, "it sank, but not as quickly as the clay shaped into a ball" and in the third column, "the long shape made the clay appear lighter than when it was in a ball shape."

Similarly, students would benefit from seeing a completed graph with the masses and volumes plotted. In this way, students know where they are headed in the lesson and have a clear understanding of the teacher's expectations.

While walking around the room questioning students about their thinking on the lesson's topics, the teacher can talk to students and glance at their written records. This gives two ways to check students' understandings without using additional class time.

When teaching lessons to ELLs, it is important to show examples of what students are expected to do independently or in groups. Behavior problems are reduced and achievement increases when students know exactly what is expected.

LANGUAGE AND CONCEPT GOALS

Sheltered lessons have clearly defined language and content objectives. Sheltered instruction teachers' lesson plans incorporate objectives that reflect content standards as well as language standards. In this way, teachers consciously integrate English language development into science topics. Many science teachers may feel that English is not their subject area, but with the increasing numbers of ELLs in schools today, all teachers should be mindful of ways to incorporate English language skills into science lessons.

For example, in the lesson presented, the teacher's language goal could have been to have students practice using academic language. To meet this objective, science vocabulary words should have been introduced, written, repeated, and highlighted for students to see, giving students access to specific academic language to use in their discussions. The teacher asked students to work in pairs, practicing oral English as they discussed their experiment. It would have been more effective if, instead of simply letting students talk, the lesson was structured in a way that guaranteed students opportunities to use academic language.

Content is adapted in sheltered lessons to students' levels of English proficiency. In this lesson the teacher did many things that made the lesson meaningful for ELLs but made a mistake when students were given a reading assignment for homework, with little or no guidance. How can a student who struggles with English be expected to read unfamiliar material from a science text at home? This type of assignment is often extremely difficult for ELLs to complete and may result in frustration and failure.

ELLs need access to grade-level text because the textbook is an important part of their learning, so sheltered instruction teachers need to adapt texts and assignments so that information is accessible to their students. Dense text can be graphically depicted, outlined, or rewritten in more understandable language. Or, passages may be read aloud and paraphrased or read with a partner. In our lesson, the teacher might have taken time to read through part of the material aloud while students followed along. An outline or guided notes would have helped students focus on the important terms and concepts. As the text was read, the teacher should have paused, drawing students' attention to related parts of the outline. After becoming familiar with the material and organizing it in outline form, students could then be assigned homework to reread the material using the outline as an aid.

This type of process focuses students in the right direction in terms of the important content from the material and it teaches them a strategy for organizing dense text into a more meaningful form. As students are explicitly taught such strategies and practice them, they become more independent in using these important learning strategies. The number of ELLs in mainstream science classes continues to grow, heightening the responsibility of science educators to find ways to make the curriculum meaningful and accessible to all students in their classes. ✧

Alan Colburn (e-mail: acolburn@csulb.edu) is an assistant professor of science education and Jana Echevarria (e-mail: jechev@csulb.edu) is a professor of special education, both at California State University, 1250 Bellflower Boulevard, Long Beach, CA 90840-4509.

NOTE

This work was supported under the Education Research and Development Program, PR/Award No. R306A60001, The Center for Research on Education, Diversity, and Excellence (CREDE), as administered by the Office of Educational Research and Improvement (OERI), National Institute on the Education of At-Risk Students (NIEARS), U.S. Department of Education (USDOE). The contents, findings, and opinions expressed here are those of the authors and do not necessarily represent the positions or policies of OERI, NIEARS, or the USDOE.

REFERENCES

Echevarria, J., and A. Graves. 1998. *Sheltered Content Instruction: Teaching English Language Learners with Diverse Abilities*. Boston: Allyn and Bacon.

Faltis, C. 1993. Critical issues in the use of sheltered content instruction in high school bilingual programs. *Peabody Journal of Education* 69(1):136-151.

Lee, O., and S. Fradd. 1998. Science for all, including students from non-English-language backgrounds. *Educational Researcher* 27(4):12-21.

Short, D., and J. Echevarria. 1998. The sheltered instruction observation protocol: A tool for teacher-researcher collaboration and professional development. Paper presented at the Annual Meeting of the American Educational Research Association, San Diego.

Science as a Second Language

Verbal interactive strategies help English language learners develop academic vocabulary

IN A SEVENTH GRADE SCIENCE CLASS, GROUPS of students are about to start an experiment. They are given small containers with water, straws, and various soap solutions and experiment with them by making bubbles. Through this experimentation, they try to find out why some bubbles burst as soon as they are blown while others float for a long time. The teacher suggests that students blow a bubble until it bursts, measure the circle produced by the bursting bubble, repeat the procedure several times, record their findings, compare the measurements recorded by other students, think of possible explanations for different findings, and discuss possible hypotheses (adapted from Cantoni-Harvey, 1987).

As is common in classrooms in the United States, this class has students whose first language is not English. Among these students, the level of English proficiency varies greatly. One student, Maria, is a recent arrival to the United States. In addition to her regular classes, she receives English as a Second Language (ESL) instruction three times a week for 45 minutes. Two other students in the class have passed the English proficiency test used in

CARMEN SIMICH-DUDGEON AND JOY EGBERT

the school district and have recently been mainstreamed into all-English instruction classes.

The teacher is certain that Maria will be unable to participate in the science experiment, so instead of including her in the science activities he has planned, he sits her close to his desk and motions for her to practice the English alphabet on a worksheet. The other English language learners, Nalan and Mehmet, both born in Turkey, are placed in separate groups so they will not use their native language. After the class, the teacher wonders if he could have integrated Maria, Nalan, and Mehmet in the science activities so that they could develop scientific knowledge and academic language at the same time.

One of the best ways to teach the content and language of science to all students, including English language learners, is to engage them in activities that promote verbal interaction and collaboration, or "interactive science teaching." In this classroom scenario, immigrant students such as Maria can and should be involved in laboratory activities early on in the English learning process. To accomplish this, the following recommendations can be adopted by middle and high school science teachers who have English language learners in their classrooms and who may or may not have support from an ESL teacher.

LANGUAGE AND CONTENT

Research suggests that both English language learners and native-English-speaking students learn science, its rhetoric, and its vocabulary best when teachers create activities and situations that allow them to be genuinely engaged in scientific inquiry through collaborative talk (Simich-Dudgeon, 1998). In addition, the use of familiar genres such as personal narratives, storytelling, and role-playing can be an effective link between English language learners' emergent English proficiency and the distinctive rhetoric and vocabulary of science.

Even science teachers who do not have specialized ESL training can develop awareness about the differences between the language and content demands of the science curriculum and the language and content knowledge of their students. English language learners can differ in their level of mathematical knowledge, which is important to consider because experimentation requires a good understanding of mathematics skills like counting, measuring, comparing, estimating, approximating, and solving equations (Cantoni-Harvey, 1987).

In addition, depending on their educational backgrounds, English language learners may lack study skills and may have low conceptual scientific understanding (Chamot and O'Malley, 1994). Experimentation, inquiry, and teacher modeling are effective tools for use with all students, particularly English language learners, because these techniques encourage them to negotiate scientific meanings through verbal interaction with the teacher and their peers.

Cultural differences may also lead English language learners to produce alternative hypotheses and interpretations than those expected by the teacher. Incorporating these students' cultural knowledge and experiences into science units and lessons makes the content easier to understand and more relevant to them. English language learners need opportunities to develop content skills such as observing, classifying, comparing, predicting, making generalizations from findings, and formulating and testing diverse hypotheses (Cantoni-Harvey, 1987).

Science teachers should be aware that English language learners, particularly those with no previous schooling or interrupted schooling, will have difficulty with the discourse, text structure, functions, and extensive vocabulary of science. Although these students may have developed the ability to communicate socially with peers and others in their homes and communities, their academic language skills may be far below grade level.

This lack of academic language is further challenged by science texts and tasks. This may be in part because science texts make it difficult for English language learners to identify the facts. For example, science texts develop concepts and skills through the use of argumentative, procedural, and descriptive genres and use different fonts, font sizes, colors, pictures, and graphic organizers to signal the organization and the importance of concepts and skills. Because these graphic elements involve so many signals, they can be confusing for those not used to them. In addition, the grammatical structure of scientific texts—frequent use of passive voice, sentences with multiple embeddings of dependent and independent clauses, complex noun phrases and structures like "if . . . then . . ." that indicate causality—may be difficult for students who are learning English (Chamot and O'Malley, 1994). Verbal interactive activities can help these students and their teachers overcome these obstacles and plan and participate in effective science learning experiences.

VERBAL INTERACTIVE ACTIVITIES

Face-to-face verbal interactions in the science classroom offer English language learners opportunities to negotiate science meanings; moreover, research suggests that talk is "a major means by which learners explore the relationship between what they already know and new observations or interpretations which they meet" (Cullinan, 1993, 2). Unlike English-speaking students who already possess valuable knowledge about the English language and its use, English language learners have English as "both a target and the medium of education" (Gibbons, 1998, 99). To facilitate the learning of English, these students need opportunities to interact with more competent speakers of English, such as the teacher and native-English-speaking peers, who act as language models for them (Simich-Dudgeon, 1998).

Verbal interactive strategies also support the *National Science Education Standards,* which emphasizes "the inclusion of all content standards in a variety of curricula that are developmentally appropriate, interesting, relevant to students' lives, organized around inquiry, and connected with other school subjects" (National Research Council, 1996, 7). These strategies allow learners to ask their own questions, incorporate personal elements into their work, and learn with and through peers. These activities can be used separately or as a whole in supporting the content and language needs of English language learners in the science classroom, and they can be adapted to fit different science curricula and texts.

INCORPORATING PERSONAL NARRATIVES

One effective verbal interactive strategy is the use of personal narratives. In this example, the teacher facilitates an inquiry project initiated by students asking about recent weather changes in their area. Based on the topic, a personal narrative activity like the one in the following scenario can help prepare students to find out more about the weather.

The teacher begins by telling a weather story from her life ("the time that the power went out during a storm" or "the hottest summer we ever had") and modeling for students the narrative genre. To help English language learners comprehend, she uses a timeline to demonstrate the process of the story and incorporates photos, magazine pictures, and information from the Internet and other family members who shared the experience. During her narrative, the teacher asks leading questions along the way (What probably happens next? How do you think it ends? Why do you say so?) and gives students time to answer. At the end of the story, she relates the story to questions about science and asks students to formulate their own questions. For example, after the power outage story she asks, "How can we be more prepared for the weather?" and "Are weather forecasts ever right?" and after the "hottest" story she asks, "Why is it so hot in the summer?" and "How can we know what the weather will be?" She posts these questions and helps students to explore ways they could find the answers to these and other questions.

Next, the teacher encourages students to develop their own weather narratives and asks leading questions of those who need help coming up with an idea or event. Students working in small groups construct timelines and other graphical supports for their narratives. Students present part of their narratives to the class and let their classmates finish the story orally or in written form, then compare their endings to what actually happened. Together, students develop questions they have about the science issues raised during the narrative segments and begin exploring these issues.

In this scenario, the teacher introduces content-based vocabulary in a personal and interesting context, and she encourages her learners to ask questions and seek answers. Students are not only relating personal narratives but also are exploring science issues with their peers through a familiar genre.

SCIENCE STORYTELLING

As part of their exploration of weather, students can develop stories using the appropriate vocabulary to describe weather processes and their importance. This procedure is illustrated in the following scenario.

A teacher elicits information from the students and models the structure of a story, from the beginning ("Once upon a time"), through the chronologically arranged body, to the conclusive end and perhaps moral ("and they lived happily ever after, but they always paid attention to the animals after that"). The students discuss the kinds of stories that might use vocabulary and ideas about

the weather. Depending on linguistic competence and science knowledge, the teacher provides story starters for students who need them.

Working in small groups, students develop their stories. Within the groups, each student has a specific role to play, such as fact checker or secretary. A group may start its story, for example, "Once upon a time there was a sad weather forecaster. He lived in San Diego where the weather never changed much. He never got to call the Weather Service Forecast Station or the TV station or use maps from the National Meteorological Center. He was very bored."

Students can post their stories around the room, read them to the class and answer questions, or compile them into a book. As students read their stories to the class, their classmates predict what comes next, provide alternative endings to the stories, and check the science facts and language for accuracy. During the story construction, all the students use the vocabulary of the content area and negotiate story content. The dual focus on science vocabulary and storytelling structure helps English language learners to be involved in interactive science.

ROLE-PLAYING IN SCIENCE

Group work is crucial for integrating English language learners into mainstream classes and supporting verbal interaction in the classroom. However, groups and group tasks must be constructed with care to ensure that individual learners have the time, opportunity, and feedback needed to participate fully in the task. Role-play can support these conditions and promote the use of content-specific oral language.

To illustrate the use of role-playing, a teacher can help his students as they work in groups to develop a script for the weather units. For example, he tells students that a hurricane (or other weather event that occurs in the students' locale) is approaching. Students assign roles and responsibilities to each group member. One role-play might feature a television weather person (who puts the science terms in lay terms), the National Weather Service Forecast Officer (who constantly updates conditions), the local weather station manager (who relays the information from the national office to local organizations such as the police, fire, and news), a local organization representative, and a concerned local citizen.

Students work together to develop a scientifically accurate conversation that requires the participation of each group member. Students perform the role-play for the class, and the teacher debriefs the class by using discussion, written reflection, or other follow-up activities.

This assignment gives learners the opportunity to personalize the science content and internalize specific vocabulary items. The teacher has the opportunity to assess learners' understanding of the concepts and the language of the lesson.

INTERACTIVE SCIENCE ACTIVITIES

In addition to the activities detailed above, other general interactive strategies have proven to be effective in supporting all students' learning of science concepts and language (Chamot,1985; Cantoni-Harvey, 1987) and are in keeping with the *National Science Education Standards*. Some of these are:

- Teachers can create Listening Centers where English language learners can listen to recorded class discussions and lessons. Students partner up to ask questions of each other and reflect on the listening task.
- During role-plays and other tasks, in addition to conveying meaning verbally, English language learners can demonstrate comprehension nonverbally through kinesics, pictures, music, visual organizers, and drawings.
- When English language learners respond to the teacher's science questions, students should be allowed to convey their thoughts as fluently as possible without focusing on their accents or grammatical errors.
- Teachers should provide ample opportunities for small-group interactions and brainstorming sessions.
- English language learners can share information with their peers by giving simple oral presentations.
- Teachers can design science Learning Centers where individuals or groups of students can engage in listening practice, science vocabulary expansion, and other content-related activities.

High school and middle school science teachers should include English language learners in interactive science learning, thereby increasing the opportunities for these students to learn science. Creating opportunities to interact within familiar genres such as personal narratives, storytelling, and role-play helps all students to learn the content and language of science. Supporting contexts that are familiar to them and tying science knowledge firmly to the students' lives provides opportunities for all students to improve their science literacy. G

Carmen Simich-Dudgeon (e-mail: csimichd@indiana.edu) and Joy Egbert (e-mail: jegbert@indiana.edu) are both assistant professors in the Department of Language Education, Indiana University, 201 N. Rose, 3044 Wright Building, Bloomington, IN 47405.

REFERENCES

Cantoni-Harvey, G. 1987. *Content-Area Language Instruction.* Reading, Mass.: Addison-Wesley.

Chamot, A. 1985. *Elementary School Science for Limited English Proficient Children. Focus 17.* Rosslyn, Va.: National Clearinghouse for Bilingual Education.

Chamot A., and J. O'Malley. 1994. *The CALLA Handbook.* Reading, Mass.: Addison-Wesley.

Cullinan, B. 1993. *Children's Voices: Talk in the Classroom.* Newark, Del.: International Reading Association.

Gibbons, P. 1998. *Learning to Learn in a Second Language.* Portsmouth, N.H.: Heinemann.

National Research Council. 1996. *National Science Education Standards.* Washington, D.C.: National Academy Press.

Simich-Dudgeon, C. 1998. Classroom strategies for encouraging collaborative discussion. In *Directions in Language and Education, No. 12.* Washington, D.C.: National Clearinghouse for Bilingual Education.

SCIENCE LEARNING FOR ALL **Celebrating Cultural Diversity**

Scientific Lit

Helping English language learners make sense of academic language

SOMETHING HAPPENED RECENTLY THAT reminded me of the importance of academic language for English language learners. My colleagues and I were dining in a restaurant when a young Vietnamese woman walked over to our table and asked if we remembered her. Several memories of her came to mind; I remembered taking blankets and dishes to her newly arrived family (when they were living in someone's basement laundry room), watching her among a group of Asian students practicing martial arts after school, and seeing her struggle determinedly with a heavy academic load.

She told us she had just graduated from the University of Maryland in chemistry, was doing summer work in a prominent research lab in our area, and had been accepted into graduate school in the fall. She then asked if her high school chemistry teacher was still at our school. She wanted to visit and thank her because she was sure that without the teacher's help and support she never could have succeeded. She was grateful to the teacher for spending so much time working one-on-one with her to polish her study skills, teaching her strategies for note taking and reading texts, and reviewing complicated vocabulary.

THE CHALLENGE OF ACADEMIC LANGUAGE

Many teachers would like to make that kind of difference in students' lives but do not think they have the time. But teachers can help English language learners in day-to-day teaching without taking a lot of extra time or providing one-on-one tutoring. Teachers should be especially aware of the challenges that language minority students face in tackling the academic language inherent in science. Language minority students are students who are considered to be fluent in English but whose first language is something else. These students may still have problems that native speakers do not. We can give these students insight into specific scientific language using tools to increase their academic language competence.

All students, including English language learners, benefit from a curriculum that emphasizes the teaching of concepts in depth and focuses on process and critical thinking skills. Minority language students can achieve higher standards of academic concepts and language if they receive support. Often, simply using the inquiry-based learning and hands-on approaches advocated by the *National Science Education Standards* (National Research Council, 1996) helps make complex rigorous academic concepts and language more concrete. But sometimes more is needed.

Academic content is especially difficult for English language learners because of the two dimensions involved: the context-reduced, or abstract, nature of most academic language and the actual complexity of the cognitive task (O'Malley and Chamot, 1990). Happily, concepts and cognitive skills can easily transfer between a student's first language and English. Students who received a good education in their own countries are more successful here academically than students whose educations have been interrupted. The successful students have developed a degree of academic language in their native tongue and often have learned some of the academic content previously. Therefore, they are not required to acquire both dimensions simultaneously. However, for many English language learners, it is likely that both the content and academic language being introduced are new to them.

TEACHING SCIENCE TERMINOLOGY

Most students, used to a context-embedded or concrete learning situation, need careful coaching in the context-reduced or more abstract scientific way of knowing embodied by the scientific inquiry process. Open inquiry incorporates six activities (Pizzini and Shepardson, 1991) that each require the ability to use specialized functional language:

■ Identification of problems and solutions and the testing of these solutions;
■ Design of procedures and data analyses;
■ Formulation of new questions based on previous claims and solutions;
■ Development of questions based on prior knowledge;
■ Linking of experience to activities, science concepts, and science principles; and
■ Sharing and discussing of procedures, products, and solutions.

Other functions students perform that require them to use academic language include seeking information,

CYNTHIA CARLSON

All students, including English language learners, benefit from a curriculum that emphasizes the teaching of concepts in depth and focuses on process and critical thinking skills.

formulating hypotheses, using time and spatial relationships, interpreting data, and so forth. Students who have had experience using these functions in a previous academic setting will be able to transfer their skills to the second language context. Students whose educational experience does not include practice in these skills, or students with interrupted education, need specific, directed training in their use.

Other features of acquiring scientific language that are especially difficult for English language learners include the use of interlocking definitions, technical taxonomies, special expressions, lexical density, syntactic ambiguity, grammatical metaphor, and semantic discontinuity (Solomon and Rhodes, 1995). Put simply, the language of science can be confusing. Words that mean one thing in general use take on whole new meanings in science class. Examples are the words culture, solution, conductor, mass, and matter. In science labs the passive voice is often used, which makes it difficult for some students to figure out who or what is doing the action described. People disappear, and objects appear to suddenly become animate. The sentence "The beaker was placed on the hot plate" may leave some students wondering who put it there. Phrases such as "the water took on a cloudy appearance" may seem to portray objects as carrying out activities.

ACCESSIBLE LANGUAGE

Teachers can help students acquire academic scientific language in several ways. Graphic organizers are wonderful tools for organizing concepts, thoughts, processes, vocabulary, reports, data, and other information in a visual, concise, and useful manner. Most science teachers already use graphic organizers such as experimental design charts, Venn diagrams, flow charts, and data tables. Consistent use of these tools enhances understanding for English language learners; the visual nature of these organizers allows for recognition of the process and/or function better than chunks of incomprehensible text or lengthy teacher explanations. When students begin to associate a process skill with a particular graphic organizer, they can begin thinking about the task, not concentrating on sorting through the language of a set of directions. Students know they will be asked to design an experiment, for example, when they see copies of an experimental design chart being distributed.

Some additional ideas for using graphic organizers include KWL charts, video logs, concept webs, and reading logs (Figure 1). Providing a blank outline for student note taking with the main headings and topics filled in provides a bit of context and takes some of the pressure off writing so students can listen more effectively. Chart formats such as "cause and effect" and "solution and consequences" can actually speed students' understanding of a particular concept.

Teachers can also provide vocabulary support through the use of word banks and word walls. Word walls are especially useful when they are divided into two sections—content words and process or functional words. Vocabulary words pertaining to a particular unit can be written on large cards beforehand or as they come up during a lesson. The words are then posted on the wall, and when teachers use the words, they point or refer to the word on the wall, thereby reinforcing the vocabulary in both an aural and visual manner. The key is this repeated connection; if the words just hang there, this technique is not as effective. The students can later use the word wall to help with spelling and memory recall during lab reports, projects, or even on tests and quizzes. As always, teachers should use real-life objects and other materials and visuals to provide a scaffold for students' understanding of material.

The use of pre-writing and reading activities provides students with the structure they need for academic tasks (Short, 1993). Teachers should familiarize students with the structure of the textbook at the beginning of the year; discuss the hierarchy of the headings and chapters; help them locate the glossary, index, and table of con-

FIGURE 1.

Graphic organizers that can be adapted well for science tasks.

KWL Chart

What we already Know about _____:	What we Want to learn about _____:	What we Learned about _____:

This chart can be used to introduce an activity or new topic by accessing background knowledge. The first two columns can be filled in individually, in groups, or as a whole class. The last column can be filled in at the end of an activity to provide closure, review, or as an informal assessment.

Video Log

What I already know about _____:	Questions I still have about _____:	Questions answered by the video:

This chart is a variation of the KWL, which is designed for use with videos.

Concept Web

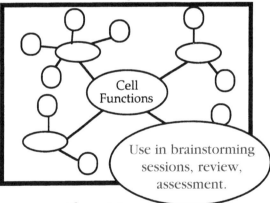

Use in brainstorming sessions, review, assessment.

Reading (or Learning) Log

What I understand about _____:	Questions I still have about _____:

This chart can be used to assess understanding of text or concepts learned in class. Can be a regular component of science journals.

Class Notes

Topic:		Date:
Subtopic 1:	1 _____ 2 _____ 3 _____	
Subtopic 2:	1 _____ 2 _____ 3 _____	
Subtopic 3:	1 _____ 2 _____ 3 _____	

Teacher can leave as is or pre-fill in topics and subtopics.

tents; and show the location of any special features of the text such as bolded vocabulary and career or technology features. Teachers should discuss each reading assignment and highlight difficult vocabulary for students before they begin. Before students begin a writing assignment, they should create an outline from which to work. Teachers could start by pre-producing an outline and gradually withdraw support over the course of the year.

One useful cooperative strategy that can make texts or articles with difficult vocabulary accessible is called "jigsaw" (Kagan, 1994). It involves forming two groupings of students: the home group and the expert group. Students begin in home groups of four or five for the introduction to the lesson. The groups then separate, and each member moves to one of four or five expert groups where they read, study, and discuss a certain text passage, article, or topic. Then, students return to the home groups, and each member either teaches the others what was learned in the expert groups or uses particular new knowledge or skills to help with a project or solve a problem. Jigsaw reading can be as simple as dividing a reading passage into four parts, or the teacher can locate reading passages with varying reading levels to allow English language learners and other students with lower reading levels to participate fully. Even if the teacher does not incorporate different reading levels, students will gain the knowledge they need to bring back to their home groups through the discussion in the expert groups.

Again, it is not the technical terms that make academic language so difficult for new speakers of English; those can be easily memorized and are usually already thoroughly covered in class because they are equally new to the native English-speaking students. The two most difficult types of vocabulary English language learners encounter are complicated functional words and common English words that are used differently in science.

Most teachers agree that the learning styles of students should also be considered; that students' talents and multiple intelligences should be given opportunities for development; and that students' cultural backgrounds and different ways of knowing should be taken into account and respected (Atwater, 1996; Gardner, 1983). This is a daunting task! The key is adding variety, not quantity. Most researchers agree that an eclectic approach to teaching is best. If teachers employ some of the approaches and techniques mentioned here and strive for a variety of activity types throughout the lesson or unit, they will become better at meeting the multiple styles and needs of students.

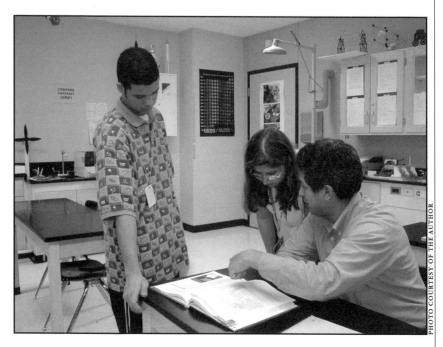

Finally, teachers should naturally incorporate a multicultural perspective in the classroom by modeling an interest and respect for other cultures, other ways of thinking, and other viewpoints. ✧

Cynthia Carlson is a sheltered science and math teacher at Montgomery Blair High School, 51 University Boulevard East, Silver Spring, MD 20912; e-mail: Cynthia_L._Carlson@fc.mcps.k12.md.us.

REFERENCES

Atwater, M.M. 1996. The multicultural science classroom: Part II: Assisting all students with science acquisition. *The Science Teacher* 63(2):20–23.

Gardner, H. 1983. *Frames of Mind: The Theory of Multiple Intelligences*. New York: Basic Books.

Kagan, S. 1994. *Cooperative Learning*. San Juan Capistrano, Calif.: Kagan Cooperative Learning.

National Research Council. 1996. *National Science Education Standards*. Washington D.C.: National Academy Press.

O'Malley, J.M., and A.U. Chamot. 1990. *Learning Strategies in Second Language Acquisition*. Cambridge: Cambridge University Press.

Pizzini, E.L., and D.P. Shepardson. 1991. Student questioning in the presence of the teacher during problem solving in science. *School Science and Mathematics* 91(8):348–352.

Short, D. 1993. Assessing integrated language and content instruction. *TESOL Quarterly* 27(4):627–656.

Solomon, J., and N. Rhodes. 1995. *Conceptualizing Academic Language: Research Report 15*: Washington, D.C.: The National Center for Research on Cultural Diversity and Second Language Learning.

Appendix A: Author Contact Information

The articles in *Science Learning for All: Celebrating Cultural Diversity* originally appeared in the journal *The Science Teacher*, 1996–2001. The following is a list of current author contact information.

S. Wali Abdi
Associate Professor of Science Education
Instruction & Curriculum Leadership
406 Ball Hall
The University of Memphis
Memphis, TN 38152
swabdi@memphis.edu

Konstantinos Alexakos
Physics Teacher
F.H. LaGuardia High School
100 Amsterdam Avenue
New York, NY 10023
kalexakos@yahoo.com

Lenola Allen-Sommerville
Assistant to the Dean
College of Education
Iowa State University
N165B Lagomarcino Hall
Ames, IA 50011
alsomm@iastate.edu

Mary Monroe Atwater
Professor of Science Education
The University of Georgia
212 Aderhold Hall
Athens, GA 30602
matwater@coe.uga.edu

Bambi Bailey
Assistant Professor of Curriculum and
    Instruction
Texas A&M International University
5201 University Boulevard
Laredo, TX 78041
bbailey@tamiu.edu

H. Prentice Baptiste
Professor, Science and Multicultural
    Education
Department of Curriculum and
    Instruction
College of Education
New Mexico State University
Las Cruces, NM 88003
baptiste@nmsu.edu

Charlotte Behm
Director, Connections Across Cultures
    Project
Jefferson Middle School
1650 West 22nd Avenue
Eugene, OR 97405
behm@eug4ja.lane.edu

Elizabeth Bernhardt
Director, Stanford Language Center
Professor of German Studies
Stanford Language Center
Building 30
Stanford University
Stanford, CA 94305
ebernhar@leland.stanford.edu

Barbara Borleske
Chemistry Teacher
John Dickinson High School
1801 Milltown Road
Wilmington, DE 19808
blborleske@redclay.k12.de.us

Melody L. Brown
Professor
College of Education
The University of Georgia
Aderhold Hall
Athens, GA 30602
mbrown@coe.uga.edu

Napoleon A. Bryant, Jr.
Professor Emeritus of Education
Xavier University
3527 Skyview Lane
Cincinnati, OH 45213
bryantn@xu.edu

Cynthia Carlson
Science and Math Teacher
Montgomery Blair High School
51 University Boulevard East
Silver Spring, MD 20910
Cynthia_L._Carlson@fc.mcps.k12.md.us

Mary Beth Carnate
Research and Development Coordinator
Department of Adult and Community
    Education
Columbus Public Schools
100 Arcadia Avenue
Columbus, OH 43202
Carnate.1@osu.edu

Alan Colburn
Professor of Science Education
California State University-Long Beach
1250 Bellflower Boulevard
Long Beach, CA 90840
acolburn@csulb.edu

Joy R. Dillard
Science Teacher
Eudora Junior High School
P.O. Box 248
Dermitt, AR 71638
Joy2daworld@altavista.com

Jana Echevarria
Professor of Special Education
California State University-Long Beach
1250 Bellflower Boulevard
Long Beach, CA 90840
jechev@csulb.edu

Joy Egbert
Department of Teaching and Learning
College of Education
Washington State University
P.O. Box 642132
Pullman, WA 99164
egbert@wsunix.wsu.edu

Richard Frost
Clinical Psychologist
Independence High School
5175 Refugee Road
Columbus, OH 43232

Tamara Garcia-Barbosa
Director of Education and Research
One Economy Corporation
2939 Van Ness Street, NW, #717
Washington, DC 20008
tjbarbosa@1-economy.com

Susan Gleason
Chemistry Teacher
Middletown High School
120 Silver Lake Road
Middletown, DE 19709
sglea@udel.edu

Stan Hill
Coordinator of K–12 Science and NSF
    Project Director
Winston-Salem/Forsyth County Schools
P.O. Box 2513
Winston-Salem, NC 27102
shill@wsfcs.k12.nc.us

Gretchen Hirsch
Associate Editor, National Center for
    Science Teaching and Learning
The Ohio State University
1929 Kenny Road
Columbus, OH 43210

Paul B. Hounshell
Professor of Science Education
University of North Carolina-
    Chapel Hill
Peabody Hall
Chapel Hill, NC 27514
pbhounsh@email.unc.edu

Ellen Johnson
Teacher, Science Department
Wilmington Friends School
101 School Road
Wilmington, DE 19803
emjohnso@udel.edu

Shirley Gholston Key
Assistant Professor of Science Education
University of Houston-Downtown
One Main Street
Suite 601 South
Houston, TX 77002
key@dt.uh.edu

Gerry M. Madrazo, Jr.
Clinical Professor and Director
Mathematics and Science Education
 Network
The University of North Carolina-Chapel
 Hill
Campus Box #3345
Chapel Hill, NC 27599
gmadrazo@email.unc.edu

Eugenie W. Maxwell
Science Teacher
Hilliard Darby High School
5410 Victoria Park Court
Columbus, OH 43235
maxwell.44@osu.edu

Marisol Rodríguez-Muñoz
Resource Teacher
Columbus Spanish Immersion Academy
2155 Fenton Street
Columbus, OH 43224

Kathryn Scantlebury
Associate Professor of Chemistry and
 Biochemistry
University of Delaware
Newark, DE 19716
kscantle@udel.edu

Carmen Simich-Dudgeon
Associate Professor and Director, Foreign
 and Second Language Education
Language Education Department
School of Education
Indiana University
Bloomington, IN 47405
csimichd@indiana.edu

Annela Teemant
Assistant Professor
School of Education
343 McKay
Brigham Young University
Provo, UT 84602
annela_teemant@byu.edu

Barbara S. Thomson
Associate Professor of Education
Mathematics, Science, and Technology
 Education
School of Teaching and Learning
College of Education
The Ohio State University
248 Arps Hall
1945 North High Street
Columbus, OH 43210
bjsthomson@aol.com

Cindy L.F. Zacks
Biology, Field Ecology, and Botany
 Teacher
Yucca Valley High School
7600 Sage Avenue
Yucca, CA 92284
zacks.d.and.c@thegrid.net

Appendix B: Resources

The following list suggests resources for further reading. Please see specific articles for references cited in those articles.

Books

American Association for the Advancement of Science (AAAS). 1989. *Science for All Americans*. Washington, D.C.: AAAS.

American Association for the Advancement of Science (AAAS). 1993. *Benchmarks for Science Literacy*. Washington, D.C.: AAAS.

American Association for the Advancement of Science (AAAS). 2001. *Atlas for Science Literacy*. Arlington, Va.: National Science Teachers Association.

Banks, J.A. 1994. *An Introduction to Multicultural Education*. Boston: Allyn and Bacon.

Barba, R.H. 1998. *Science in the Multicultural Classroom*. Needham Heights, Mass.: Allyn and Bacon.

Bernstein, L., A. Winkler, and L. Zierdt-Warshaw. 1998. *African and African-American Women of Science*. Maywood, N.J.: Peoples Publishing Group.

Bernstein, L., A. Winkler, and L. Zierdt-Warshaw. 1998. *Latino Women of Science*. Maywood, N.J.: Peoples Publishing Group.

Bernstein, L., A. Winkler, and L. Zierdt-Warshaw. 1996. *Multicultural Women of Science*. Maywood, N.J.: Peoples Publishing Group.

Doran, R., F. Chan, and P. Tamir. 1998. *Science Educator's Guide to Assessment*. Arlington, Va.: National Science Teachers Association.

Epps, C.H., D.G. Johnson, and A.L. Vaughan. 1994. *African-American Medical Pioneers*. Baltimore: Williams and Wilkins.

Gibson, J.T. 1999. *Developing Strategies and Practices for Culturally Diverse Classrooms*. Norwood, Mass.: Christopher Gordon Publishers.

Golemba, B.E. 1992. *Lesser-Known Women: A Biographical Dictionary*. Boulder: Lynne Reiner Publishers, Inc.

Grant, C.A., and G. Ladson-Billings, eds. 1997. *Dictionary of Multicultural Education*. Phoenix: Oryx Press.

Haber, L. 1970. *Black Pioneers of Science and Invention*. New York: Harcourt, Brace, & World.

Kessler, J.H., J.S. Kidd, R.A. Kidd, and K.A. Morin. 1996. *Distinguished African American Scientists of the Twentieth Century*. Phoenix: Oryx Press.

Morey, A.I. and M.K. Kitano, eds. 1997. *Multicultural Course Transformation in Higher Education: A Broader Truth*. Boston: Allyn and Bacon.

National Research Council (NRC). 1996. *National Science Education Standards*. Washington, D.C.: National Academy Press.

National Science Teachers Association (NSTA). 1997. *Pathways to the Science Standards, High School Edition*. Arlington, Va.: NSTA.

National Science Teachers Association (NSTA). 2000. *Pathways to the Science Standards, Elementary School Edition.* Arlington, Va.: NSTA.

National Science Teachers Association (NSTA). 2000. *Pathways to the Science Standards, Middle School Edition.* Arlington, Va.: NSTA.

Rosenthal, J.W. 1997. Multicultural science: Focus on the biological and environmental sciences. In *Multicultural Course Transformation in Higher Education: A Broader Truth*, A.I. Morey and M.K. Kitano, eds. Boston: Allyn and Bacon.

Rhoton, J., and P. Bowers, eds. 1995. *Issues in Science Education.* Arlington, Va.: National Science Teachers Association.

Rhoton, J., and P. Bowers, eds. 2000. *Issues in Science Education: Professional Development Leadership and the Diverse Learner.* Arlington, Va.: National Science Teachers Association.

Rhoton, J., and P. Bowers, eds. 2000. *Issues in Science Education: Professional Development Planning and Design.* Arlington, Va.: National Science Teachers Association.

Van Der Does, L.Q., and R.J. Simon. 2000. *Renaissance Women in Science.* Lanham, N.Y.: University Press of America.

Warren, W. 1999. *Black Women Scientists in the United States.* Bloomington, Ind.: Indiana University Press.

Wilson, M., and E. Snapp, eds. 1992. *Options for Girls: A Door to the Future.* Austin: PRO-ED.

Wodowski, R.L., and M.B. Ginsberg. 1995. *Diversity and Motivation—Culturally Responsive Teaching.* San Francisco: Jossey Bass.

Articles

Allen, G.G., and O. Seumptewa. 1988. The need for strengthening Native American science and mathematics education. *Journal of College Science Teaching* 18(5):364–369.

Atwater, M.M. 1993. Multicultural science education. *The Science Teacher* 60(3):33-37 (reprinted in 67(1):48-49).

Barba, R.H., V.O. Pang, and M.T. Tran. 1992. Who really discovered aspirin? *The Science Teacher* 59(5):26–27.

Blake, S. 1993. Are you turning female and minority students away from science? *Science and Children* 30(7):32–35

Bryant, N. 1988. Sons, daughters, where are your books? *Journal of College Science Teaching* 18(5):364–369

Clark, J.V. 1988. Black women in science: Implications for improved participation. *Journal of College Science Teaching* 18(5):348–352

Connor, J.M. 2000. Studying racial bias: Too hot to handle? *Journal of College Science Teaching* 30(1):26-32.

Dooley, E., G. Bardwell, and C. Bethea. 2000. Mentors in medicine. *The Science Teacher* 67(3):36-39.

Hays, E.T. 2001. Defining multicultural science education. *The Science Teacher* 68(2):8.

Hrabrowski, F.A., III, and W. Pearson, Jr. 1993. Recruiting and retaining talented African-American males in college science and engineering. *Journal of College Science Teaching* 22(4):234–238

Madrazo, G. and P.B. Hounshell. 1993. Multicultural education: Implications for science education and supervision. *Science Educator* 2(1):17-20.

Mason, C.L., and R.H. Barba. 1992. Equal opportunity science. *The Science Teacher* 59(5):22–26.

Mittag, K.C., and D. Mason. 1999. Cultural factors in science education—variables affecting achievement. *Journal of College Science Teaching* 28(5):307-310.

Monhardt, R.M. 2000. Fair play in science education: Equal opportunities for minority students. *The Clearing House* 74(1):18–22.

Rakow, S.J., and A. Bermudez. 1988. Underrepresentation of Hispanic Americans in science. *Journal of College Science Teaching* 18(5):353-355.

Reichert, B. 1990. Not all of those giants were European. *Science Scope* 13(5):47-49.

Selin, H. 1993. Science across cultures, part I. *The Science Teacher* 60(3):38–44

Selin, H. 1993. Science across cultures, part II. *The Science Teacher* 60(4):32–36.

Shih, F.H. 1988. Asian American students: The myth of a model minority. *Journal of College Science Teaching* 18(5):356-359.

Smith, N.C. 2000. Empowering African Americans in the sciences. *Journal of College Science Teaching* 30(3):156-157.

Tippins, D.J., and N.F. Dana. 1992. Culturally relevant alternative assessment. *Science Scope* 15(6):50-53.

Websites

ASPIRA Association, Inc.
www.aspira.org

Eisenhower National Clearinghouse (ENC)
www.enc.org

Washington University's ESL on the Internet
www.artsci.wustl.edu/~langlab/depts/esl/

The Faces of Science: African Americans in the Sciences
www.princeton.edu/~mcbrown/display/faces.html

The Internet Teachers of English as a Second Language (TESL) Journal
www.aitech.ac.jp/~iteslj

Latino Education Directory
www.ael.org/eric/maed

Teacher of English to Speakers of Other Languages (TESOL)
www.tesol.edu

Urban Education Web
eric-web.tc.columbia.edu

Women in Math
darkwing.uoregon.edu/~wmnmath